T0231038

The Genetics
of Social Evolution

Westview Studies in Insect Biology
Michael D. Breed, Series Editor

The Genetics of Social Evolution, edited by Michael D. Breed and Robert E. Page, Jr.

The Entomology of Indigenous and Naturalized Systems in Agriculture, edited by Marvin K. Harris and Charlie E. Rogers

Interindividual Behavioral Variability in Social Insects, edited by Robert L. Jeanne

Fire Ants and Leaf-Cutting Ants: Biology and Management, edited by Clifford S. Lofgren and Robert K. Vander Meer

Management of Pests and Pesticides: Farmers' Perceptions and Practices, edited by Joyce Tait and Banpot Napompeth

Integrated Pest Management on Rangeland: A Shortgrass Prairie Perspective, edited by John L. Capinera

The "African" Honey Bee, edited by David J.C. Fletcher and Michael D. Breed (forthcoming)

The Genetics of Social Evolution

EDITED BY

Michael D. Breed
and Robert E. Page, Jr.

Routledge
Taylor & Francis Group

LONDON AND NEW YORK

First published 1989 by Westview Press

Published 2019 by Routledge
52 Vanderbilt Avenue, New York, NY 10017
2 Park Square, Milton Park, Abingdon, Oxon OX14 4RN

Routledge is an imprint of the Taylor & Francis Group, an informa business

Copyright © 1989 Taylor & Francis

All rights reserved. No part of this book may be reprinted or reproduced or utilised in any form or by any electronic, mechanical, or other means, now known or hereafter invented, including photocopying and recording, or in any information storage or retrieval system, without permission in writing from the publishers.

Notice:
Product or corporate names may be trademarks or registered trademarks, and are used only for identification and explanation without intent to infringe.

Library of Congress Cataloging-in-Publication Data
The Genetics of social evolution / edited by Michael D. Breed and
     Robert E. Page, Jr.
        p. cm.—(Westview studies in insect biology)
     1. Insects—Genetics. 2. Insects—Evolution. 3. Insects—
Behavior. 4. Insect societies. 5. Social behavior in animals.
6. Behavior evolution. I. Breed, Michael D. II. Page, Robert E., Jr.
III. Series.
QL493.G46 1989
595.7'0524—dc19   88-12229
                                                                                           CIP

ISBN 13: 978-0-367-29248-5 (hbk)
ISBN 13: 978-0-367-30794-3 (pbk)

CONTENTS

PREFACE

Our primary goal in organizing this book was to initiate a synthesis of thought on how genetics structures the behavior of individual animals that live within complex social systems. To do this we have brought together leading theorists and empiricists who apply genetics to the study of eusocial insect evolution.

The eusocial insects, with differentiated worker and reproductive castes, pose particularly important problems in evolutionary genetics. The most apparent of these is the question of how extreme altruism (sterile worker castes) might have evolved. Twenty-five years later, the elegant genetic kin selection arguments of Hamilton still lack conclusive tests. Strassmann and Queller (Chapter 5) and Queller and Strassmann (Chapter 6) consider the current status of kin selection explanations. In doing so they focus on the relative importance of ecology and genetics in understanding the evolution and maintenance of extreme altruism. *Polistes* wasps are the source of their empirical data. Kukuk (Chapter 10) presents genetic data from populations of the primitively eusocial sweat bee, *Lasioglossum zephyrum* (also refered to as *Dialictus zephyrus*), and discusses the application of these data to kin selection hypotheses. Kin recognition may be a prerequisite or a logical consequence of kin selection; in Chapter 9, Mintzer presents genetic models for the kin recognition system of *Acacia* ants.

Selection may act on a social group as a unit. Owen (Chapter 3) considers the conditions under which the colony can be considered a unit of selection and discusses the genetic consequences of colony-level selection.

Empirical studies have found considerable genetic variation within colonies of many species, despite the genetic cohesion of eusocial groups that is normally assumed. Page et al. (Chapter 2) and Robinson and Page (Chapter 4) look at intracolonial genetic structure in honey bees and its impact on division of labor. Ross (Chapter 8) documents the genetic structure of polygynous ant colonies and presents data on differential reproductive output of queens.

Social parasitism is particularly problematical. Social parasites are often considered to be sister species to their hosts. If this is the case, then special mechanisms of speciation might be invoked to explain host/parasite divergence. Ward (Chapter 7) considers the mechanisms by which social parasites have evolved and their role in shaping social diversity.

The new conceptual bases of thought in evolutionary genetics, as presented in the chapters in this book, are largely dependent on a new array of techniques. Included in the new arsenal of evolutionary genetics are computer modeling, biochemical techniques that allow access to different genetic markers, and more sophisticated modes of ecological and behavioral analysis. Further advances on the questions raised here will depend on the application of these techniques to a set of taxa that allows a broadly comparative understanding of eusocial systems.

Most of the papers in this book were first given in a symposium at the National Conference of the Entomological Society of America, December 1987. We are grateful to the Society for its financial support of the symposium. We would also like to express our thanks to Westview Press, and particularly Kellie Masterson, for advice and encouragement of this project. Finally, we are indebted to the Psychobiology Program of the National Science Foundation for its support of our research programs.

M. D. B. and R. E. P.

CHAPTER 1

INTRODUCTION: THEORY AND EMPIRICAL RESULTS IN SOCIOGENETICS

Michael D. Breed

The domain of sociogenetics includes karyotype evolution, sex determination and mating systems, genetics of social behavior, kin recognition, and caste and levels of selection (Crozier 1987). Eusocial insects are particularly fertile ground for studies in sociogenetics; they have complex behavioral patterns which are often poorly understood from an evolutionary point of view and they are compatible with a variety of genetic approaches. The purpose of this volume is to present a broad view of current work that relates genetics to the special evolutionary features of ants, bees, and wasps.

In this chapter I define some terminology that will be used throughout the book. I also discuss some of the major conceptual issues that are addressed in later chapters and give some views on some of the evolutionary processes leading to the origin and maintenance of sterile castes. Sociogenetics is an emerging area of research; some of the major issues in the sociogenetics of eusocial insects are discussed.

LEVELS OF EUSOCIALITY

Batra (1966) introduced the term **eusocial** to describe colonies of insects in which there is a reproductive division of labor. Eusocial colonies have three defining features: 1.) castes and division of labor, 2.) matrifilial colony organization with two generations of adults present, and 3.) cooperative brood care. In such colonies some second generation individuals have reproductive capacity reduced or absent and raise the offspring of the parental generation. Individuals with reduced fecundity are called **workers** and the fecund parent is the **queen**.

Other possible social organizations include: **communal**,

Department of Environmental, Population, and Organismic Biology, The University of Colorado, Boulder, CO USA 80309-0334 individuals share a nest but each rears its own brood; **semisocial**, the female parent is not present but there is a division of reproduction and labor among the remaining sisters in the colony; and **quasisocial**, the mother and/or her offspring share a nest and cooperate in nest construction but there is no division of reproduction (Michener 1974).

All termites and ants, wasps in the family Vespidae, a small number of wasps in the family Sphecidae, and many species of bee are eusocial. In eusocial ants, bees, and wasps colonial organization is truly matrifilial, with the daughters of the queen serving as workers. Males are usually present only as ephemeral guests in the colony prior to mating. In termites, colonies are usually founded by a male and female pair, the **king** and queen, and both male and female offspring serve as workers.

In wasps and bees there is a gradation among species in caste differentiation between the queen and workers. Species, such as the sweat bee *Lasioglossum (Dialictus) zephyrum* (see Kukuk, chapter 10, this volume), have little or no morphological differentiation between the queen and her workers and are called **primitively eusocial**. Species such as the honey bee, *Apis mellifera*, with striking queen-worker differentiation, are called **highly eusocial**. All ants and termites are highly eusocial. Eusocial insects are sometimes informally referred to as "social insects".

The eusocial Hymenoptera--ants, bees, and wasps--have a **haplodiploid** mode of sex determination. Normally, fertilized, diploid, eggs develop into females while unfertilized, haploid, eggs produce males. The sperm from any one male carry identical genomes, barring spontaneous mutations. This identity of sperm genomes results in high levels of relatedness among females that share the same parents (super sisters--see below). Termites are diplodiploid, although Lacy (1980, but see Crozier and Luykx 1985) has proposed special genetic mechanisms that could make termites functionally haplodiploid organisms.

POLYGYNY AND POLYANDRY

In some cases colony genetic structure is complicated by the presence of multiple queens or by multiply-mated queens. In the social insect literature colonies with multiple queens are considered to be **polygynous**. This definition of polygyny differs from the convention in the mammalian and avian literature (e.g., Emlen and Oring, 1977) and readers must be careful not to confuse the two usages.   Single queen colonies are called **monogynous**.

In ants, bees, and wasps the queen mates in the early stages of her adult life, often prior to colony foundation, and does not mate again.   Sperm is stored from this mating and is used to fertilize female eggs (unfertilized eggs develop into males).   If the queen mates more than once during this early mating period she is termed **polyandrous**. The most extreme documented example of social insect polyandry is the honey bee; queens may store sperm from between 15 and 20 matings (Page 1986).   Sperm depletion in queen honey bees may be pathological to the colony, as male eggs will be laid in worker comb, but honey bee queens do not seek new matings in response to such depletion.   In contrast termite queens mate repeatedly through their life.

Polygyny and polyandry are acutely important because each may increase the genetic diversity in the colony and reduce the average relatedness among workers.   Diversity and relatedness have important impacts on evolutionary models; these effects are discussed in this chapter, and in subsequent chapters in this volume.

WITHIN FAMILY RELATIONSHIPS

Polygyny and polyandry create complex genealogical relationships in colonies which overwhelm the traditional terminology for describing such relationships.   Page and Laidlaw (1988) proposed a new descriptive system, which is summarized here.   The basic problem is that the term "full sister" has been used in diplodiploid systems to describe individuals with average $r=0.50$ and in haplodiploid systems to describe individuals with an average $r=0.75$ (assuming no inbreeding).   Page and Laidlaw (1988) propose that **super sister** be used when comparing individuals that share a diploid mother and a haploid father.   This terminology focuses on the origins of genomes involved in zygote formation and acknowledges the higher average degree of relatedness among the offspring of such a mating due to the sperm being identical.   Using the same genetic pairing terminology, female progeny that have the same mother and the same paternal grandmother are **full sisters**; the average r among such individuals is 0.50.   **Maternal half sisters** have the same mother and unrelated fathers; the average $r=0.25$.   The terminology of Page and Laidlaw (1988) is used throughout this book.

3

INTRACOLONIAL GENETIC VARIATION

One of the most important sociogenetic issues derives from genetic variation in eusocial insect colonies. The workers in eusocial insect colonies are not genetic clones. Empirical and theoretical studies of the evolution of insect societies often use colony-average relatedness values. They thereby assume that each nestmate is of equal genetic "value" to another or that interactions are purely stochastic in nature, e.g., individuals are treated alike because no one can tell the difference! In fact, chromosomal segregation can lead to a wide range of worker relatednesses even if the queen has mated only once. Patterns of competition and cooperation among workers should be structured in correspondence with these genetic differences and similarities.

In the extreme case, polygyny and/or polyandry in a colony generate even greater differences among workers, including the possibility that some workers in the same colony are not related at all. Polygygny and polyandry are frequent in the eusocial insects (e.g. West-Eberhard 1987b) and provide useful systems in which to study the behavioral implications of genetic differences within colonies. Page *et al.* (Chapter 2, this volume) and Robinson and Page (Chapter 4, this volume) use genetic differences due to polyandry as a tool to initiate studies in this area. Ross (Chapter 8, this volume) presents data that illustrate some of the implications of polygyny.

Robinson and Page (1988) and Frumhoff and Baker (1988) have presented convincing evidence that genetic differences among subfamilies in honey bee colonies are significant in structuring the pattern of labor in those colonies. This is, in my view, only the tip of a very large and interesting iceberg of instances in which the detailed genetic structure of colony populations plays a large role in colony biology.

Consideration of the complexity of colony structure leads to the next area of consideration, the bases for the evolution of sterile castes.

STERILE CASTES

Two Possible Evolutionary Mechanisms

Explaining the evolution of sterile or only partially fecund workers has been recognized as a major evolutionary problem

4

since Darwin's time (see Page *et al.*, Chapter 2 this volume). In simple form, there are two possible solutions to this problem. First, workers may reproduce indirectly by adding to the reproductive success of their more fecund relatives. Kin selection of this sort was dramatically and thoroughly explicated by Hamilton (1964) and has been supported by a number of subsequent authors (e.g. West-Eberhard 1975, Trivers and Hare 1976, Oster and Wilson 1978, Craig 1983). Alternatively, workers may serve as phenotypic extensions of their mother and may, because of pre- or post-eclosion parental manipulation not be, as individuals, evolutionary units (Michener and Brothers 1974, Alexander 1974). Crozier (1982) gave a insightful review of thought in this area.

Perhaps Hamilton's (1964) most important insight was that the hymenopteran haplodiploid sex determination system results in asymmetries of relationship within colonies. Assuming no inbreeding, monogyny, and single mating, all the workers in the colony are super sisters and have average relatednesses of 0.75. The mother-daughter relationship, whether it is between the queen and her workers or between the workers and daughters that they might have, is 0.50. Thus workers are actually more related to their sisters than their daughters; this increases the probability of transmission of genes shared among sisters and synergizes the kin selection effects of aid given to the queen by workers.

Testing Kin Selection

As Strassmann and Queller (Chapter 5, this volume) point out, there has been a distressing paucity of empirical tests to determine if kin selection plays an important role in the maintenance of worker behavior. After more than ten years of studies measuring relatednesses (beginning with Metcalf and Whitt 1977) the only thing that is clear is that genetic relationships among nestmates within insect societies are not sufficiently high to explain the differential probability of eusocial evolution between haplodiploid and diploid species.

Tests that have been conducted are of two sorts. Trivers and Hare (1976) used comparative data to test a series of predictions. These had different outcomes depending on whether kin selection and manipulation arguments were hypothesized. Their principal prediction was that if kin selection is the driving factor in social evolution then the sexual investment ratio of the colony should be under worker control and approximate 3:1. If parental control is the driving factor then the sexual investment ratio should approximate 1:1. Data that they collected from a large number of taxa seems to support a kin selection argument, although Alexander and Sherman (1977) criticized many of the assumptions made by Trivers and Hare (1976). Noonan (1978) found a 1:1 sex investment ratio in *Polistes* wasps and Page and Metcalf (1984) argued that the sex investment ratio in the honey bee approximates 1:1. These additional tests of this hypothesis seem to support a manipulation argument, although the social system in honey bees is so highly evolved that sex ratio may have been altered secondarily to worker loss of fecundity.

There are two significant problems with using sex ratios to test for kin selection. First, it is usually unclear what hypothesis is being tested. Kin selection is one of many factors that may drive sex ratios away from 1:1. Second, the empirical data are nearly impossible to get. Investment is difficult to define and measure. It is equally difficult to account for the different genetic factors that may operate on sex determination.

The other way in which kin selection has been tested is by using electrophoretically detectable markers to trace genealogies or to estimate relatednesses among individuals in colonies. The reproductive benefit of aiding behavior can then be estimated by comparing colonies of different sizes or with different patterns of aid-giving. Queller and Strassmann (Chapter 6, this volume) review the few studies in which this has actually been done and suggest methods for interpreting electrophoretic data. Kukuk (Chapter 10, this volume) uses this approach in studies of a primitively eusocial bee. Not nearly enough studies of this sort have been performed to draw broad conclusions concerning evolutionary mechanisms.

Kin Recognition as a Test of Kin Selection

Indirect support for kin selection comes from demonstrations of kin recognition mechanisms. In order for an animal to behave differentially towards relatives it must have some way of identifying which conspecifics are closely related and which are not. In some cases the spatial location of an individual (e.g., co-membership in a litter) confers that information. In most cases, however, olfactory, auditory, or visual cues are used by animals to discriminate close relatives from other individuals (reviews: Fletcher and Michener 1987). Although many animal species are known to be able to discriminate close kin from other conspecifics, much work now needs to be done to test if discrminations can be made between different types of relatives (e.g., between cousins and brothers). Differential behavior correlated with such discriminations can be strong support for kin selection arguments.

Testing Manipulation

Direct tests of manipulation arguments have been even rarer. The existence of dominance hierarchies has been interpreted as support for manipulation arguments (Michener and Brothers 1974). In some species the nutrition of worker larvae is under queen control and queens may repress workers by nutritionally limiting their fecundity (Craig 1983).

In fact the two main hypotheses, kin selection and parental manipulation, are by no means mutually exclusive. To my knowledge, no one has credibly challenged the basic logic of kin selection, as Hamilton (1964) formulated it. Similarly, queens are well endowed with opportunities to manipulate workers (Alexander 1974) and there is no logical basis for excluding the possibility of manipulation. Any factor of an individual's biology that reduces the probability of it establishing an independent nest, or which reduces its potential fecundity, increases the potential relative benefits to that individual of serving as a worker. Thus parental manipulation of worker reproductive potential changes the kin selection cost/benefit ratio for the worker. Thus it makes no sense to present the arguments as dichotomous hypotheses (West-Eberhard 1975), as some authors seem to have done (Alexander 1974, Trivers and Hare 1976). This view is supported by the modelling efforts of Craig (1979) who showed that manipulation and kin selection cannot be treated separately.

Mutualism

Lin and Michener (1972) and West-Eberhard (1978a) show that under certain circumstances mutualism can explain apparently altruistic behavior. West-Eberhard (1978a) found evidence in a wasp, *Metapolybia*, that individual queens work for mutual benefit during certain parts of the colony cycle and compete for reproductive status later in the cycle. Mutualistic explanations for cooperative behavior are attractive when the short-term costs of helping are outweighed by possible long-term

7

payoffs.

Ecological Factors

Lin (1964), West-Eberhard (1975), Evans (1977) and Strassmann and Queller (Chapter 5, this volume) point out that ecological factors can play a key role in the cost/benefit equation as well. Strassmann and Queller, quite rightly, stress that this is perhaps the most ignored feature of the entire scheme and argue very cogently that further studies should focus on the ecological costs and benefits of different strategies of helping at the nest versus founding separate nests.

Ultimately, however, the question of sterile caste establishment and maintenance may be a genetic question (but see the arguments of West-Eberhard 1987a 1987b). How do genes that promote worker sterility and cooperation (whether they be genes controlling a manipulative phenotype in the mother or genes controlling an "altruistic" phenotype in the daughter) proliferate over time in a population (Craig 1979)? Fully successful genetic models, which have not yet been produced, will consider the relative contributions of maternal and offspring genotypes to worker behavior, interactions among workers that are consistent with genetic variation in the worker population, and ecological constraints on workers and queens. None of these factors work in a vacuum, and they all impact the selective processes.

Epigenetic Explanations

Another avenue by which empirical support for theories explaining sterile castes is obtained are searches for sets of species that represent stages of social evolution. Michener (1985) argued that the evolutionary transition between solitary and eusocial may be abrupt, rather than going through such a series:

> ...while the scenarios usually cited, involving a series of species intermediate between the solitary and the eusocial conditions, may in some cases be correct, they are sometimes wrong. There is increasing evidence that the primitively eusocial way of life can arise...without intervening *species* that are semisocial, subsocial, etc. The potential exists for direct development of eusociality involving individuals within a hitherto solitary bee population. (Michener 1985, p. 294.)

8

His argument has been extended by West-Eberhard (1987a 1987b), who suggests that it may be misleading to focus on allelic models of social evolution. She points out that many of the features of eusocial behavior are already present within the range of possible phenotypic flexibility observed in solitary species. This epigenetic theory posits that social characteristics may appear via pleiotropic effects of facultative traits selected in solitary insects or via contextual shifts in the expression of a facultative trait. However, the work of Robinson and Page (1988) and Frumhoff and Baker (1988) raises questions about whether behavior that is apparently facultative is truly so.

LEVELS OF SELECTION

The view of a eusocial colony as a superorganism has had its ups and downs (see Page *et al.*, this volume). One fault with this analogy was that it led to a predominating view of the colony as the unit upon which selection acted (Wilson 1971 reviews this type of thinking). The entrance of Hamilton's (1964) work into the arena forced the pendulum in the opposite direction; eusocial colonies came to be viewed as aggregations of phenotypically cooperating but genetically competing individuals, each with its own selective imperatives.

Selection acts on the queen when the colony dies to the extent that a colony is the phenotypic extension of the queen (see Owen this volume). In many or most cases, however, workers have some individual reproductive capacity. Examples of this include male-destined eggs produced by workers in many species of ants (Trivers and Hare 1976) and in the honey bee (Page and Erickson 1988). Strongly expressed reproductive behavior in workers may disrupt cooperative brood rearing, foraging, or other essential activities in a colony and result in death of the entire group. Thus while individual selection may favor deviations from patterns of individual altruism, colony-level selection will oppose extreme deviations.

COLONY ORGANIZATION

Our understanding of how social organization within colonies has evolved is even less well developed. Oster and Wilson (1978) attempted to model caste differentiation. Their models match the number of morphological groups in colonies with the number of tasks requiring distinctly different morphologies. A number of interesting questions stem from their work. Of particular interest are the ecological factors that underly different caste structures and the ontogenetic stages that a colony undergoes as it increases in size. Other than Wilson's work on *Atta* (1983) surprisingly little work was stimulated by Oster and Wilson's (1978) book. If we are to fully understand eusocial evolution then the interplay between the evolution of worker infertility and the evolution of the worker's enhanced ability to work needs to be clarified. This is clearly an important area for further studies.

CONCLUSIONS

    ·   The crux of explaining the derivation of sterile castes, and of explaining the complexity of existing eusocial systems, is understanding why and how individuals bias their behavior. Social interactions in colonies of eusocial insects will only be fully understood if the genetic similarities/differences among the individual actors are known. Individual variation in behavior is an essential focus for future studies on eusocial insects (Jeanne 1988).

Kin-biased behavior, as an explanation for individual variation, has been a convenient hypothesis; extensive arguments have been built that these biases play an important role in social evolution. However, there have been few empirical demonstrations of situations in which assessment of kinship is linked to a decision concerning what behavior to express. Detailed knowledge of genetic similarities within a social unit may provide experimental tools which can be used in finally testing this basic hypothesis.

Few authors have grappled with the fact that the relationships among individuals in a eusocial colony of Hymenoptera, even one in which the queen is singly mated, are highly variable. Kin selection arguments, in their extreme, predict affiliations among individuals that are explained by this variation. Polygyny and polyandry amplify the variation and provide convenient ways for exploring fine grained genetic differences.

In sum, sociogenetics provides worthwhile tools and a useful conceptual basis for developing a better understanding of behavioral mechanisms in eusocial colonies. Perhaps most important is an appreciation for the fact that variation, the "grist of evolution, in behavior is as important as the mean, and that testing hypotheses concerning correlations between genetic variation and behavioral variation should be fruitful.

LITERATURE CITED

Alexander, R. D. 1974. The evolution of social behavior. *Ann. Rev. Ecol. Syst.* 5:325-383.

Alexander, R. D. and P. W. Sherman. 1977. Local mate competition and parental investment in social insects. *Science* 196:494-500.

Batra, S. W. T. 1966. Life cycle and behavior of the primitively social bee, *Lasioglossum zephyrum. Univ. Kansas Sci. Bull.* 46:359-423.

Craig, R. 1979. Parental manipulation, kin selection, and the evolution of altruism. *Evolution* 33:319-334.

Craig, R. 1983. Subfertility and the evolution of eusociality by kin selection. *J. Theor. Biol* 100:379-397.

Crozier, R. H. 1982. On insects and insects: Twists and turns in our understanding of the evolution of eusociality. In *The Biology of Social Insects.* M. D. Breed, C. D. Michener, and H. E. Evans, eds. Westview Press:Boulder, Colo. pp. 4-9.

Crozier, R. H. 1987. Towards a sociogenetics of social insects. In *Chemistry and Biology of Social Insects*, J. Eder and H. Rembold, eds. Verlag J. Peperny:Munich. pp. 325-328.

Crozier, R. H., and P. Luykx. 1985. The evolution of termite eusociality is unlikely to have been based on male-haploid analogy. *Am. Nat.* 126:867-869.

Emlen, S. T. and L. W. Oring. 1977. Ecology, sexual selection, and the evolution of mating systems. *Science* 197:215-223.

Fletcher, D. J. C. and C. D. Michener. 1987. *Kin recognition in animals.* John Wiley & Sons: New York.

Frumhoff, P. C. and J. Baker. 1988. A genetic component to division of labour within honey bee colonies. *Nature* 333:358-361.

Evans, H. E. 1977. Extrinsic and intrinsic factors in the evolution of insect sociality. *Bioscience* 27:613-617.

Hamilton, W. D. 1964. The genetical evolution of social behavior, I and II. *J. Theor. Biol.* 7:1-52.

Jeanne, R. L. 1988. *Interindividual behavioral variability in social insects.* Westview Press:Boulder, Colo.

Lacy, R. C. 1980. The evolution of eusociality in termites: a haplodiploid analogy? *Am. Nat.* 116: 449-451.

Lin, N. 1964. Increased parasitic pressure as a major factor in the evolution of social behavior in halictine bees. *Insectes Sociaux*

11:187-192.

Lin, N. and C. D. Michener. 1972. Evolution of sociality in insects. *Quart. Rev. Biol.* 47:131-159.

Metcalf, R. A. and G. S. Whitt. 1977. Relative inclusive fitness in the social wasp, *Polistes metricus. Behav. Ecol. Sociobiol.* 2:353-360.

Michener, C. D. 1974. *The comparative social behavior of the bees.* Belknap Press of Harvard University Press:Cambridge, Mass.

Michener, C. D. 1985. From solitary to eusocial: need there be a series of intervening species? In *Experimental Behavioral Ecology*, B. Holldobler and M. Lindauer, eds. G. Fischer Verlag:Stuttgart. pp. 293-305.

Michener, C. D. and D. J. Brothers. 1974. Were workers of eusocial Hymenoptera initially altruistic or oppressed? *Proc. Natl. Acad. Sci. USA* 71:671-674.

Noonan, K. M. 1978. Sex ratio of parental investment in colonies of the social wasp, *Polistes fuscatus. Science* 199:1354-1356.

Oster, G.F., and E.O. Wilson. 1978. *Caste and ecology in the social insects.* Princeton University Press:Princeton.

Page, R. E. 1986. Sperm utilization in social insects. *Ann. Rev. Entomol.* 31:297-320.

Page, R. E. and H. H. Laidlaw. 1988. Full sisters and super sisters: a terminological paradigm. *Anim. Behav.* 36:944-945.

Page, R. E. and R. A. Metcalf. 1984. A population investment sex ratio for the honey bee (*Apis mellifera* L.). *Am. Nat.* 124:680-702.

Page, R. E. and E. H. Erickson. 1988. Reproduction by worker honey bees. *Behav. Ecol. Sociobiol.* 23:117-126.

Robinson, G.E., and R.E. Page. 1988. Genetic determination of guarding and undertaking in honey bee colonies. *Nature* 333:356-358.

Trivers, R. L. and H. Hare. 1976. Haplodiploidy and the evolution of social insects. *Science* 191:249-263.

West-Eberhard, M. J. 1975. The evolution of social behavior by kin selection. *Quart. Rev. Biol.* 50:1-33.

West-Eberhard, M. J. 1978a. Temporary queens in *Metapolybia* wasps: non-reproductive helpers without altruism? *Science* 200:441-443.

West-Eberhard, M. J. 1978b. Polygyny and the evolution of social behavior in the wasps. *J. Kansas Entomol. Soc.* 41:832-856.

West-Eberhard, M. J. 1987a. The epigenetical origins of insect sociality. In *Chemistry and Biology of Social Insects*, J. Eder and H. Rembold, eds. Verlag J. Peperny:Munich. pp. 369-371.

West-Eberhard, M. J. 1987b. Flexible strategy and social evolution. In *Animal Societies: Theories and Facts*, Y. Ito, J. L. Brown, and

J. Kikhawa, eds. Japan Sci. Soc. Press:Tokyo. pp. 35-51.

Wilson, E. O. 1971. *The Insect Societies*. Belknap Press of Harvard University Press:Cambridge, Mass.

Wilson, E. O. 1983. Caste and division of labor in leaf cutter ants IV. Colony ontogeny in *A. cephalotes*. *Behav. Ecol. Sociobiol.* 14:55-60.

CHAPTER 2

GENETIC STRUCTURE, DIVISION OF LABOR, AND THE EVOLUTION OF INSECT SOCIETIES

Robert E. Page, Jr., Gene E. Robinson, Nicholas W. Calderone, and Walter C. Rothenbuhler

Social insects presented Charles Darwin with major difficulties for his fledgling theory of evolution by natural selection. Sterile individuals in insect societies seemed to violate the very basic principle of his theory that the process of adaptation is based on the differential survival and reproductive success of individuals. The fact that sterile workers are often differentiated from the reproductives, both physically and behaviorally, added a further complication, the "acme" of which, to Darwin, was the observation that sterile workers are even further differentiated into separate physical and behavioral subgroups (Darwin 1962, pp. 268-273).

Darwin resolved his difficulties by invoking a level of selection that he called selection on families, now called colony-level selection (Crozier and Consul 1976; and Owen 1986). He proposed that the process of natural selection acting at the level of the colony could result in both the evolution of a sterile group (the worker caste) and in physical and behavioral differences within this caste, leading to the formation of distinct subcastes. The "grist" of this evolutionary process is individual variability among workers that results in differences in colony reproductive success.

A modern interpretation of Darwin's model should be based on three processes: 1. behavior of individual workers is affected by genotype--the worker's own, that of the queen (assumed to be related), or those of related worker nestmates, 2. the genotypically based differences in behavior lead to differences in colony performance that differentially affect reproductive components of fitness on which colony-level selection acts; and 3. colony-level selection changes the genotypic distribution of workers and reproductives. This explanation for the evolution of social organization focuses on evolutionary processes. These depend on the expression of variablity in individuals and is conceptually different from other explanations of social organization (Wilson 1971 1985a, Oster and Wilson 1978).

Oster and Wilson (1978) found evidence for the role of natural selection in the evolution of social organization by studying organizational properties of colonies. This phenomenological approach successfully describes the division of labor found in many insect societies. Individual specialization on specific tasks, age-related changes in behavior (age or temporal polyethism), and the differentiation of groups of individuals into physically and behaviorally distinct classes that perform more-or-less discrete sets of tasks (physical- and age-castes; see Wilson 1976; Jeanne 1986) have been compared to similar properties of metazoan organisms (Wilson 1985a,b). The resultant "superorganism" metaphor suggests that insect societies are adapted to maximize their ergonomic efficiency by responding to environmental change with behavioral, physiological, and sometimes anatomical differentiation of individuals.

Environmental effects on the differentiation of individuals have been studied and support this view. However, as a consequence of polyandry, polygyny, and recombination, individuals within colonies are not genetically identical, as are most cells of a metazoan. Genotypic heterogeneity may result in genetic constraints on the behavioral plasticity of individuals and may affect organizational properties of colonies. These potential genetic effects on colony organization are important for understanding the evolution of insect societies as proposed by Darwin and are intractable with a phenomenological approach that ignores genetic differences among individuals.

In this paper we show that processes resulting in changes in the frequencies of worker genotypes can account for the evolution of the organization of honey bee societies. We also develop a general model of division of labor that incorporates this genetic level of organization and discuss the possible adaptive significance and maintenance of intranidal genetic variability.

GENETIC DETERMINATION OF INDIVIDUAL BEHAVIOR

Social insect workers display a finite repertoire of behavioral acts, each one of which is highly stereotyped, typical of the species, and genetically programmed (Wilson 1971; Wilson and Oster 1978). However, not all individuals perform the same sets of tasks with the same frequencies. Inter-individual variability in behavior has been previously assumed to be primarily, if not exclusively, the result of an individual's membership in different temporal or physical subcastes, a consequence of the current "needs" of the colony as determined by the colony environment. Recent evidence for honey bees now shows that much of the observed variability in worker behavior is a consequence of genetic variability for two components of division of labor: task partitioning--the probability of performing a specific behavioral act from the repertoire of all behavioral acts--and age polyethism.

Robinson and Page (1988, see also Chapter 4 this volume) established colonies of honey bees with three biochemically-distinguishable subfamilies (individuals of the same subfamily share the same mother and father, those of different subfamilies have different fathers). Members of different subfamilies engaged in different tasks with different probabilities. These results demonstrate the presence of genetic variability for performing specific tasks. In addition, the genotype of a worker affects her own behavior; this is the first essential Darwinian process for the evolution of division of labor.

The ability of colony-level selection to change worker behavior can be demonstrated by analogy with studies employing artificial selection. Hellmich *et al.* (1985) initiated a two-way selection program for pollen hoarding. Queens were selected as virgin-queen mothers and sources of males (males are haploid and produced by parthenogenesis) on the basis of the quantity of pollen stored in combs of the nest. Four generations of artificial, colony-level selection resulted in two distinct subpopulations: one that hoarded large quantities of pollen, and one that hoarded very little.

Calderone and Page (1988) studied the behavior of individual workers from these two subpopulations to determine the effect of colony environment on pollen collecting. Workers from the two lines were raised in the same colony at the same time, and were transferred into the same observation hives that contained unselected bees. Even though they had identical rearing environments and were members of the same colony throughout their adult lives, they behaved very differently. Workers from the high-pollen hoarding line maintained their pollen collecting behavior while those of the low-hoarding line did much less pollen collecting (FIGURE 1). They also began foraging at different ages showing a genotypic effect on age polyethism. Bees of the lines differed with respect to where they were located within the nest

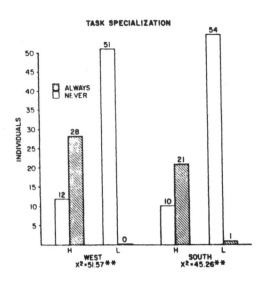

FIGURE 1. The number of individuals belonging to the high- (H) and low- (L) pollen hoarding lines that were observed by Calderone and Page (1988) in two colonies (West and South) that either always or never returned from a foraging trip with a load of pollen. Only bees that were observed more than once are included. ** = difference between high and low lines was significant (P<0.01) based on chi-square contingency analysis.

prior to the onset of foraging, suggesting that they were performing tasks at different frequencies. A follow-up experiment using the same methodology demonstrated that same-aged individuals of the different lines did differ in their frequencies of performing different within-nest tasks (Calderone and Page, unpublished data).

Rothenbuhler (unpublished data) conducted a similar study in 1969 that showed genetic variability for the performance of specific tasks that are normally associated with different-aged bees. He established colonies with 200 young bees (less than 2 days old) from each of two different sources. In one colony, 200 bees came from a colony of his "Brown line" and 200 from his "Van Scoy line". The Brown line was characterized by its hygienic behavior (workers uncap cells and remove dead larvae) and strong colony defense, while the Van Scoy line did not show hygienic behavior and was not defensive (for more information on hygienic behavior see Rothenbuhler 1964a, 1964b). The hive entrance opened into a large cage that restricted the flight activity of the

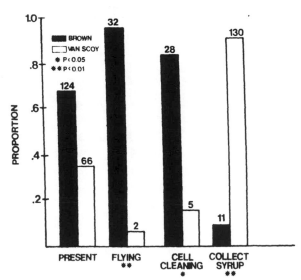

FIGURE 2. The proportion of bees belonging to the Brown and Van Scoy lines that were observed by Rothenbuhler (unpublished data) present in the hive, flying off the combs during inspection (considered a defensive response), cleaning cells containing dead larvae, and collecting syrup from a feeder. Numbers at the tops of bars show the total number of observations for that category. Data were analyzed by contingency table.

workers. A feeding station with sugar syrup solution and pollen was provided. About 4 weeks later, he took a census of surviving, marked bees of the initial cohort and made a series of behavioral observations. His results show that the genotype of an individual influences the work it does (FIGURE 2). Identically-aged but genotypically different individuals distributed themselves differently among tasks. The tasks performed reflected the characteristics of the lines to which they belonged: Brown-line workers performed primarily preforaging-age tasks while Van Scoy bees foraged.

Collectively, these results demonstrate the second and third elements of Darwin's model: 2. the behavior of individual workers leads to colony traits on which colony-level selection can act, and 3. colony-level selection results in the change in individual behavior of workers.

The demonstration of genetic variability for components of division of labor may lead to a more complete model of the evolution of colony organization that includes the effect of individual genotypes. In the following section we present a heuristic model based on genetic studies of honey bees in an attempt to construct a testable, conceptual framework for studying the evolution of division of labor.

DIVISION OF LABOR MODEL

The principal objective for a model of division of labor should be to determine the likelihood that an individual observed during any instant in time will perform a specific task. The collection of all probabilities for all individuals for all possible tasks should then predict the activities of a colony. Assume that $P(T)_k$ is the probability that individual k performs task T. Then

$$P(T)_k = \mu + S_i + A_{ij} + e_{ijk} \quad (1)$$

In this case μ is a populational parameter representing the average probability that any individual in the population will perform task T; S_i is the increment (positive or negative) to this probability that is a consequence of individual k's membership in subfamily i within a given colony; A_{ij} is the increment associated with individual k belonging to age class j within subfamily i (age class can be defined for any interval of age); and e_{ijk} is the residual deviation that includes effects of the environment-- including the colony environment, unspecified interactions, and other random effects.

The S_i term represents genetic contributions to the variance in the probability of performing a task. It is composed of additive and non-additive components from the genomic contributions of the queen and drone father of any given subfamily. If age does not affect probability, e.g., no temporal subcastes, then all variance that is strictly a consequence of an individual's genotype is included in this term.

If, however, age affects probability, then some genetic components could be included in the A_{ij} term. This term represents the results of selection for the temporal-caste structure of colonies and can be expanded to

$$A_{ij} = \alpha_j + SA_{ij} + e_{ij} . \quad (2)$$

Here, α is a populational parameter representing the average

20

probability that an individual of age j will perform task T--a consequence of age polyethism. SA_{ij} increments or decrements as a consequence of belonging to age-class j within subfamily i, and represents genetic variability for the rate at which individuals change age-related behavior, their sensitivity to environmental conditions that result in induced changes in behavior, and/or the degree of discretization of behavior according to temporal-caste membership (Wilson 1976; Seeley 1982). The residual effect is e_{ij}, a component the residual term e_{ijk} in (1), that may be associated with environmental conditions unique to members of age class j, subfamily i. Such conditions may arise through non-random associations of colony members due to kin recognition or other mechanisms.

With this model, the allocation of work effort in a colony at any given time will be the expected number of individuals in that colony that are engaged in all possible tasks, including non-tasks. The number of individuals expected to engage in task T is:

$$E(N_T) = N\Sigma_i\Sigma_j\Sigma_k \ P(T)_{ijk} \quad (3)$$

where N = the total number of individuals in the colony.

Based on the above model and discussion (see FIGURE 3), an individual honey bee worker progresses through a series of age castes (the number of age castes recognized varies with different investigators, see Seeley 1982; cf Kolmes 1986). Within a given age caste, individuals are likely to perform a more-or-less discrete set of tasks (Wilson 1976; Seeley 1982). The probability that an individual will perform a given task within a task set is determined by both genetic and environmental effects. As discussed above, qualitative evidence exists for genotypic components of variance for both the task partitioning, S_i, and the age polyethism, SA_{ij}, components of division of labor (Calderone and Page 1988, Robinson and Page 1988; see also Chapter 4, this volume). In addition, strong environmental influence, confounded in the e_{ijk} term, has been qualitatively determined (see Ribbands 1953, pp. 307-309) However, quantitative estimates of these components are not available.

THE ADAPTIVE SIGNIFICANCE OF GENETIC VARIABILITY

We have shown above (see also chapter 4) that considerable genetic variability exists for both the task

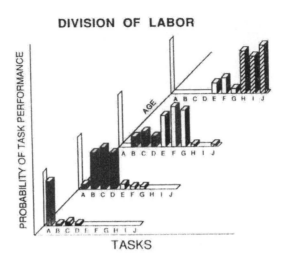

FIGURE 3. Diagram of model of division of labor that is discussed in text.

partitioning and age polyethism components of division of labor. If these elements of division of labor are important, then why hasn't the variability been reduced to less easily detectable levels by directional selection for optimal values? Current models for the maintenance of genetic variability in social insect populations may be classified as follows.

The variability itself is unimportant to individual colonies; but the traits are. Genetic variability may be maintained by directional selection in a spatially or temporally heterogeneous environment (Hedrick 1986). In that case the variability itself is accidental but the trait is important. Variability exists because the environment is not constant enough for natural selection to reach an "optimal" solution and eliminate the variability.

Behavioral dominance (see Robinson and Page, Chapter 4, this volume) could affect both the rate at which directional selection reduces genetic variability and the amount of variability maintained at mutation-selection equilibrium (Crow and Kimura 1970). Behavioral dominance results in a reduction in selection response and may lead to even higher levels of genetic variability in heterogeneous environments.

22

The variability and the traits are both unimportant: It could also be that the traits involved are themselves relatively unimportant and not subject to strong directional selection. In that case, directional selection would be less able to eliminate genetic variability in finite populations subject to genetic drift. Expected levels of genetic variability would also be greater at mutation-selection equilibrium in very large populations (Crow and Kimura 1970).

The variability and the traits are important: Several models have been proposed for the selective advantages of genetic heterogeneity for worker traits within insect societies. Crozier and Consul (1976) demonstrated that genetic polymorphism can be maintained at a single locus by colony-level selection if heterozygous workers make colonies more fit than do homozygous workers.

Owen (1986; see also Chapter 3 this volume) demonstrated the conditions for maintaining genetic polymorphism at a single locus with colony-level selection in populations with once and twice-mated queens. His models investigated the effects of both additive and non-additive interactions between workers of different genotypes. In general, genetic polymorphism is maintained in populations under colony-level selection if genotypic interactions are additive and heterozygous workers result in higher colony fitness than homozygotes. However, if genotype interactions are non-additive, then the magnitude and sign of the fitness effects of the interactions will determine the equilibrium gene frequencies and if any polymorphisms are maintained in the population.

Moritz and Southwick (1987) also stressed the importance of genotypic interactions in determining colony phenotype. They considered positive, non-additive, interactions to be analogous to overdominance in classical genetics. They suggest that mechanisms that support genotypic diversity (such as polygyny and polyandry) should be favored by natural selection.

Genetic variability could be maintained if it increases the frequency of "genetic specialists" in colonies. These hypothetical individuals have genotypes that are relatively rare and result in a greater degree of specialization on specific tasks (Robinson and Page 1988, see also Chapter 4, this volume). Specialization then may result in greater efficiency of individuals and increase colony reproductive output. We present this model below.

For this model, we assume that two loci interact to determine the probability that an individual worker will perform one of two alternative tasks. For illustration, these may be the two principal foraging activities of honey bees: pollen and nectar collecting. These assumed loci are dosage compensated so that all individuals forage with the same expected frequency and success; only the probability of foraging for pollen or nectar is affected. Locus A controls the likelihood of foraging for pollen and locus B the likelihood of foraging for nectar. An individual AAbb forages exclusively for pollen; aaBB exclusively for nectar; AaBb, AABB, and aabb forage 50% for pollen; AABb and Aabb 75% for pollen; and AaBB 75% for nectar. Assume that the optimum labor allocation for a colony is for 50% of the workers to forage for pollen and 50% for nectar. Colonies that have worker genotypes that result in a 50:50 allocation are most fit and are assigned a fitness of 1.0.

There are several ways a colony can have the optimum allocation: it can have all double heterozygotes, AaBb; it can have all double homozygotes AABB and aabb; or, it can have any one of several equal mixes of genotypes such as AAbb and aaBB, AaBB and AABb, etc. Colonies that allocate all of their labor to either pollen or nectar collecting are least fit and are assigned a fitness value of 0.0. Fitness is linearized from 50:50 to 0:100 and 100:0. For example, 75:25 lies midway between 50:50 and 100:0 and has an associated fitness of 0.50 as does a 25:75 allocation.

Genetic specialists are individuals whose genotypes specify a probability of foraging for pollen or nectar that deviates from 50:50. The degree of specialization is determined by the degree of deviation. For example, an AAbb individual is exclusively a pollen forager and is a complete specialist, degree of specialization = 1.00. An Aabb individual is a "half specialist", degree of specialization = 0.50, etc. The proportion of specialists in a colony is the weighted average of the degree of specialization of all genotypes.

Specialists may increase colony fitness by increasing reproductive output through greater efficiency. A colony that is composed of equal numbers of AAbb and aaBB genotypes may have a higher level of productivity, through increased specialization of colony members, than a colony consisting of all AaBb genotypes, even though both colonies allocate work optimally. Now, colony fitness can be defined by the following relationship:

24

$$W_C = W_A (1 + I) \qquad (4)$$

where: W_C = colony fitness;

W_A = fitness based on allocation ratio;

I = the average degree of specialization (S) X the coefficient of specialization (σ).

The coefficient of specialization determines the degree of increase in fitness as a consequence of specialization. A coefficient of $\sigma = 0$ means there is no increase in fitness associated with specialization and $W_C = W_A$. A coefficient of $\sigma = 1$ means that a colony composed entirely of specialists (S = 1) is twice as fit as one composed entirely of non-specialists, $W_C = 2W_A$.

FIGURE 4 shows the expected fitness of a population in Hardy-Weinberg equilibrium where each queen mates one time. Fitness values are a function of the frequencies of the A and B alleles for the three cases where $\sigma = 0$, 1.5, and 3. Genetic polymorphism can be maintained in a population only if genetic specialists increase colony fitness. When the coefficient of specialization is 0 (no increase in fitness) the optimum gene frequencies lie at the opposing corners of the fitness surface fixed for either the "A" and "B" or the "a" and "b" alleles (top of Figure 4). Fixation of these alleles eliminates specialist genotypes. A minimum threshold value of $\sigma = 1.6$ (see middle, FIGURE 4) is necessary before any fitness values lie above those at the highest corners (fitnesses of 1.00). When σ is 1.6 or greater (see bottom, FIGURE 4), then a ridge of polymorphic gene frequencies with values lying on the diagonal between (0,0) and (1,1) will be favored, hence maintaining genetic polymorphism and genetic specialists.

Polyandry relaxes the conditions for maintaining a polymorphic population. A minimum value of $\sigma = 1.1$ is sufficient for populations where queens mate twice. In addition, polyandry results in higher average fitnesses of populations in which σ is equal to or greater than 1.1, and, therefore, may be favored by selection. Measures of the changes in efficiency of workers with increasing levels of specialization are needed to assess the value of

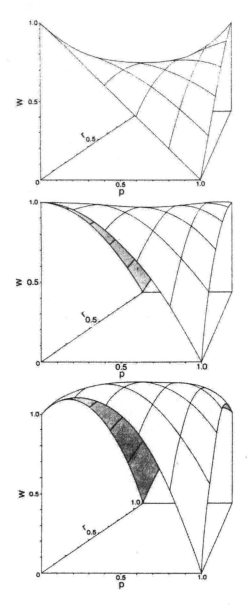

FIGURE 4. Diagram of the three-dimensional surface of fitness values for the model presented. W = fitness, p = frequency of the "A" allele, r = frequency of the "B" allele. Shaded area is the underside of the surface. σ= 0, 1.5, and 3.0 for the top, middle, and bottom, respectively.

the model presented here.

CONCLUSION

The investigation of the genetic bases of division of labor in insect societies has only just begun. Using the honey bee as a model system, significant genetic components of inter-individual variability in behavior have been demonstrated. The results of artificial colony-level selection demonstrate by analogy the Darwinian processes involved.

Genetic models that demonstrate the conditions necessary to maintain genetic variability for individual behavior now become important for understanding the adaptive significance, if any, of the observed variability. Through these models and the results of empirical studies, the genotypes of individuals are becoming important factors in studying the organization of insect societies because genotype determines both the probability that individuals of a given age caste perform specific tasks, and determines the rate of behavioral ontogeny. Whether genotypic variability within nests resulting from genetic recombination or polyandry is adaptive remains to be demonstrated. However, genotypic variability does represent a level of social organization.

The remaining challenge is to demonstrate how genotype and environment act in concert to integrate social behavior. The study of mechanisms of behavior and genetic structure of colonies may add to our understanding. A preliminary model that integrates mechanistic behavior and genetics has already provided testable hypotheses (Robinson and Page 1988; see also Robinson and Page Chapter 4 this volume). The ultimate goal is to demonstrate how genetically and behaviorally variable individuals live in societies that themselves exhibit the properties of a superorganism.

Acknowledgements

This work was funded by National Science Foundation Grants BNS-8615381 and BNS-8719283 to REP, an Ohio State University Postdoctoral Fellowship Award to GER, an Ohio State University Presidential Fellowship Award to NWC, and Public Health Service Research Grant HD-00368-04 to WCR.

LITERATURE CITED

Calderone, N. W. and R. E. Page. 1988. Genotypic variability in age polyethism and task specialization in the honey bee, *Apis mellifera* (Hymenoptera: Apidae). *Behav. Ecol. Sociobiol.* 22:17-25.

Crow, J. F. and M. Kimura. 1970. *An introduction to population genetics theory.* Burgess Publishing Co.:Minneapolis.

Crozier, R. H. and P. C. Consul. 1976. Conditions for genetic polymorphism in social Hymenoptera under selection at the colony level. *Theor. Pop. Biol.* 10:1-9.

Darwin. C. R. 1962. *The origin of species by means of natural selection or the preservation of favoured races in the struggle for life, 6th edition.* Macmillan Publishing Co., Inc.:New York.

Hedrick, P. W. 1986. Genetic polymorphism in heterogeneous environments: a decade later. *Ann. Rev. Ecol. Syst.* 17:535-566.

Hellmich, R. L., J. M. Kulincevic and W. C. Rothenbuhler. 1985. Selection for high and low pollen-hoarding honey bees. *J. Hered.* 76:155-158.

Jeanne, R. L. 1986. The evolution of the organization of work in social insects. *Monitore Zool. Ital. (N.S.)* 20:119-133.

Kolmes, S. A. 1986. Age polyethism in worker honey bees. *Ethology* 71:252-255.

Moritz, R. F. A. and E. E. Southwick. 1987. Phenotype interactions in group behavior of honey bee workers (*Apis mellifera* L.). *Behav. Ecol. Sociobiol.* 21:53-57.

Oster, G. F. and E. O. Wilson. 1978. *Caste and ecology in the social insects.* Princeton University Press:Princeton.

Owen, R. E. 1986. Colony-level selection in the social insects: single locus additive and nonadditive models. *Theor. Pop. Biol.* 29:198-234.

Ribbands, C. R. 1953. *The behaviour and social life of honeybees.* Bee Research Association Ltd.:London.

Robinson, G. E. and R.E. Page. 1988. Genetic determination of guarding and undertaking in honey bee colonies. *Nature* 333:356-358.

Rothenbuhler, W. C. 1964a. Behaviour genetics of nest cleaning in honey bees. I. Response of four inbred lines to disease-killed brood. *Anim. Behav.* 12:578-583.

Rothenbuhler, W. C. 1964b. Behaviour genetics of nest cleaning in honey bees. IV. Responses of F_1 and backcross generations to disease-killed brood. *Am. Zool.* 4:111-123.

Seeley, T. D. 1982. Adaptive significance of the age polyethism schedule in honeybee colonies. *Behav. Ecol. Sociobiol.* 11:287-293.

Seeley, T. D. 1986. Division of labor among worker honeybees. *Ethology* 71:249-251.

Wilson, E. O. 1971. *The insect societies.* Belknap Press of the Harvard University Press:Cambridge, Mass.

Wilson, E. O. 1976. Behavioral discretization and the number of castes in an ant species. *Behav. Ecol. Sociobiol.* 1:141-154.

Wilson, E. O. 1985a. The sociogenesis of insect colonies. *Science* 228:1489-1495.

Wilson, E. O. 1985b. The principles of caste evolution. In *Experimental behavioural ecology and sociobiology.*, B. Holldobler and M. Lindauer, eds. Sinauer Associates, Inc.:Sunderland pp. 307- 324.

CHAPTER 3

THE GENETICS OF COLONY-LEVEL SELECTION

Robin E. Owen

The idea that natural selection could be effective at the level of the family, or colony in the case of the social insects, originated with Charles Darwin (1860) when he wrote (p. 237) that "... selection may be applied to the family, as well as to the individual, ...". He continued, ...

> I do not doubt that a breed of cattle, always yielding oxen with extraordinary long horns, could be slowly formed by carefully watching which individual bulls and cows, when matched, produced oxen with the longest horns; and yet no one ox could ever have propagated its kind. Thus I believe it has been with social insects: a slight modification of structure or instinct, correlated with the sterile condition of certain members of the community, has been advantageous to the community: consequently the fertile males and females of the same community flourished, and transmitted to their fertile offspring a tendency to produce sterile members having the same modification (Darwin 1860 p. 238).

Note one point of clarification; Darwin, when he refers to "the community", clearly means the family, not the community of organisms as a whole (Sober 1984).

In this paper I define colony-level selection to mean that if *inherited variation of characters expressed by workers, or reproductive offspring, exists and leads to differential productivity or survival of colonies, then the reproductive success of the founding queen and male is determined by selection*

Department of Biological Sciences, University of Calgary, Calgary, Alberta, T2N 1N4 Canada *at the level of the colony.* This, I believe, is very close to what Darwin intended. Therefore, colony-level selection, as defined here, is selection on mating-types or fertilities of founding queens and males for properties of colonies, which are the traits expressed by the workers. Mating-type (or fertility) selection simply means that the reproductive success of an individual depends not only on its own genotype but also on the genotype of its mate. Mating is assumed to be at random with respect to the gene locus under consideration thereby distinguishing mating-type from sexual selection, except for a few special cases (Feldman *et al.* 1983).

Group selection is gene frequency change due to differential extinction or productivity of groups (Wade 1978), therefore colony-level selection can be considered a type of group selection in which the groups are families (colonies). However, I think that conceptually it is clearer to regard, as did Darwin (1860), colony-level selection as a form of mating-type selection. This is because the properties (phenotypes) of colonies, that are expressed through the workers, are just a secondary expression of the genotypes of the founding queen and male.

Colonies exhibit properties as a whole, not shown by individual workers themselves, which are not merely epiphenomena but represent adaptive traits in their own right (Wilson 1985). A commonly held view of a social insect colony was as a "superorganism", with the various members being analogous to the cells and organs of a metazoan body and having a similar level of organization (Starr 1979). This view gradually became unfashionable because the analogy to an individual organism became strained and difficult to maintain, while no real explanatory power emerged (Wilson 1971). Although the idea of colony-level selection has sometimes been linked with the superorganism concept (e.g. West-Eberhard 1975 1979), the models discussed in this paper do not depend on regarding the colony as a superorganism. Even properties of colonies as a whole, expressed by workers as "group" or "aggregate" characters (e.g. alarm reaction, nest thermoregulation, allometric relationships) can be analyzed using methods of quantitative genetics, and fall within the framework of mating-type selection.

CHARACTERS SELECTED AT THE LEVEL OF THE COLONY

These belong to three categories, the first two involve characters directly expressed by the workers, while the third

32

involves characters of the reproductives (queens and males).

Unitary Characters

Unitary characters are those expressed by individual workers. Social insect workers perform many varied tasks including foraging, brood rearing, nest construction and defense. If genetic variation for performance of tasks such as these exists in a population, then the genotypes of the workers will affect the survival or productivity of the colony. Similarly genetic variation for developmental flexibility could exist; one genotype might allow workers to follow different developmental pathways (depending, for example, on nutritional status), while the other genotype(s) would always produce invariant workers. Other examples include morphological traits such as color variation, allozyme variation, or tolerance to heat or cold. For a trait controlled by a single diallelic locus, colonies consist of all homozygous, all heterozygous or a mixture of homo- and heterozygous workers. Therefore, the colony phenotype and fitness depend on the proportion of each worker phenotype present. In some cases colony fitness depends on whether or not a certain task is performed and this may be completed by a small proportion of workers in the colony. Examples include necrophoric behavior (removal of dead individuals from the nest) by ants (Wilson *et al.* 1958, Howard and Tschinkel 1976) and honey bees (Rothenbuhler 1964, Visscher 1983), and location of foraging patches by independent scouting in honey bees (Seeley 1983).

Characters expressed jointly by the workers and the reproductives (either queens alone or queens and males), for instance allozyme or color variation, may be selected at both individual and colony levels. However, colony-level selection alone acts on those characters expressed exclusively by the workers.

Aggregate (Or Group) Characters

These are characters expressed only by workers collectively in a social group as a whole; it is not possible for a worker to show the character individually. A distinction can be made between behavioral and morphological characters. An example of the former is reaction to alarm pheromones by honeybees and wasps (Moritz and Bürgin 1987). The change in rate of oxygen consumption as a response to isopentyl-acetate (a major component of alarm pheromone) by honey bee workers can only be measured on groups, as solitary individuals do not respond (Southwick and Moritz 1985). A less obvious type of character, which is morphological, is the allometric relationship between body dimensions of workers in a colony. The morphological dimensions of any given worker are fixed but the worker's behavioral response depends on its interactions with other members of the group.

Characters of Reproductives

Characters expressed only by the reproductives (queens and males) may, in some instances, be selected for at the level of colonies. One example, to be discussed in detail later, is body size of reproductives. Body size could be set by the intrinsic growth rate of the genotypes of the reproductives, alternatively it could be controlled by the workers through larval feeding rates. In this case the workers' genotypes are phenotypically expressed in the reproductives. As long as there is a cost to the colony of producing reproductives varying in body size, then colony-level selection can occur.

COLONY-LEVEL SELECTION, KIN SELECTION AND FERTILITY SELECTION

Colony-level selection is a type of kin selection, because the genotypic fitness of individuals within a population are affected by their relatives. This is particularly clear if family-structured models are considered. There are six different family types possible at a single autosomal locus with two alleles (TABLE 1). Most models of kin selection examine the conditions for the increase in frequency of "altruistic" alleles -- those harmful (in terms of survival and reproduction) to the bearer but beneficial to the family group. Thus if offspring have different fitness within families (TABLE 1) then the altruistic alleles decrease in frequency within families each generation, but may still spread through the population if the overall fitness of the families containing altruists is increased sufficiently. Wade (1980) has shown explicitly that total gene frequency change per generation (Δp) under kin selection is the sum of two components, one caused by individual selection within families and the other caused by

TABLE 1. Family types and fitnesses for a model of kin selection in a family structured population (modified from Michod 1982).

Mating type	Frequency[1]	Offspring fitness within families			Mean family fitness
		A_1A_1	A_1A_2	A_2A_2	
$A_1A_1 \times A_1A_1$	D^2 p^4	W_{01}			$W_1 = W_{01}$
$A_1A_1 \times A_1A_2$	$2DH$ $4p^3q$	W_{02}	W_{12}		$W_2 = 0.5(W_{02} + W_{12})$
$A_1A_1 \times A_2A_2$	$2DR$ $2p^2q^2$		W_{13}		$W_3 = W_{13}$
$A_1A_2 \times A_1A_2$	H^2 $4p^2q^2$	W_{04}	W_{14}	W_{24}	$W_4 = 0.25W_{04} + 0.5W_{14} + 0.25W_{24}$
$A_1A_2 \times A_2A_2$	$2HR$ $4pq^3$		W_{15}	W_{25}	$W_5 = 0.5(W_{12} + W_{25})$
$A_2A_2 \times A_2A_2$	R^2 q^4			W_{26}	$W_6 = W_{26}$

[1] Hardy-Weinberg genotype proportions used in many models when selection is assumed to be weak.

selection between families. This can be written,

$$\Delta p = \Delta p_B + \overline{\Delta p W}$$

where Δp_B is change in gene frequency due to selection between family groups, and $\overline{\Delta p W}$ is the mean change in p by individual selection within families (Wade 1980). Michod (1982) points out that models of kin selection are in no way confined to the study of altruism (although kin selection may be necessary for altruism), and potentially any character could evolve through kin selection.

Colony-level selection deals with selection between families. The fitness of the colony (family in the simplest case) is assumed to be a function of the contributions of the worker genotypes present within the colony. By just considering fitness differences between families, colony-level selection is also equivalent to fertility or mating-type selection, as Darwin (1860) realized. There are a number of models applicable to fertility selection at an autosomal locus (Penrose 1949, Bodmer 1965, Cockerham *et al.* 1972, Hadeler and Liberman 1975, Roux 1977, Pollak 1978, Templeton 1979, Butler *et al.* 1982, Feldman *et al.* 1983, Feldman and Liberman 1985, Liberman and Feldman 1985). A few authors have also considered haplodiploid loci either for the case of weak selection (Nagylaki 1979 1987, Abugov 1983 1985) or particular special cases (Crozier and Consul 1976). The results of all these models are relevant to colony-level selection, but in this section I present a model of colony-level selection (Owen 1986) which is applicable to unitary characters of social Hymenoptera (haplodiploids). This brings out the differences between the outcome of individual and colony-level selection, and shows the effects of multiple-mating and worker-produced males on colony-level selection.

Colony fitness

At a single diallelic haplodiploid locus there are six different mating types (TABLE 2), giving rise to three homogenotypic colony types (those comprised of a single worker genotype (i.e. A_1A_1, A_1A_2 or A_2A_2), and two heterogenotypic colony types (those containing workers of two genotypes, i.e., A_1A_1 and A_1A_2 or A_1A_2

and A_2A_2). Colony fitness can be assigned in a number of different ways. In the most general case each colony has an independent fitness, W_i. In the special case considered by Crozier and Consul (1976) the homogenotypic and heterogenotypic colony fitnesses are independent (TABLE 2). Alternatively, colony fitnesses can be thought of as a function of the contributions made by the different worker genotypes. If colony fitness is conceptualized in this way then the fitnesses of the homogenotypic and heterogenotypic colonies are not independent. For instance, the fitnesses of colony types 1 and 3 (TABLE 2) are correlated because both contain A_1A_1 workers. The simplest way to assign colony fitness is to take the arithmetic average of the contributions made by the worker genotypes; this is the additive model. However one might expect there to be interactions in the heterogenotypic colonies which would raise or lower the colony fitness nonadditively. To model this, the fitnesses of the heterogenotypic colonies are modified by an interaction term e_i. Thus in the nonadditive model with $0 \leq e_i < 1.0$ fitnesses are less than under the additive model, while with $1.0 < e_i \leq \infty$ fitnesses are greater. For instance, if A_1A_1 and A_1A_2 workers have different ranges of cold tolerance, then foraging by heterogenotypic colonies will be possible over twice the range of temperatures compared to homogenotypic colonies, thus $e_i > 1$. When $e_1 = e_2 = 1.0$ the non-additive model reduces to the additive model. In TABLE 2 u_{ij} denotes the contribution made by the worker of genotype A_iA_j to the colony fitness. In terms of the nonadditive model the special case of colony fitnesses of Crozier and Consul (1976) corresponds to $u_{11} = u_{22} = V$, $u_{12} = 1$ and $e_1 = e_2 = e = 2x/(1+V)$.

Recurrence equations and equilibrium solution

Let D, H and R be the frequencies of the genotypes A_1A_1, A_1A_2 and A_2A_2 respectively, with $D + H + R = 1$. Random mating, discrete generations and an infinite population size are assumed. Males can have two origins; from the workers in proportion ψ and from the queen in proportion $(1 - \psi)$, where $0 \leq \psi \leq 1$. A_1 and A_2 males are at frequencies p_m and q_m respectively.

In the next generation the female genotypic frequencies will be,

TABLE 2. Haplodiploid colony types with four different colony-level selection viability schemes.

Mating type Queen x Male		Colony Fitness Worker Genotypes In Colony		General Case	Additive Model	Non-Additive Model	Special Case (Crozier & Consul, 1976)
1. A_1A_1	A_1	A_1A_1		W_1	u_{11}	u_{11}	V
2. A_1A_1	A_2	A_1A_2		W_2	u_{12}	u_{12}	1.0
3. A_1A_2	A_1	$0.5A_1A_1$		W_3	$0.5(u_{11} + u_{12})$	$0.5(u_{11} + u_{12})e_1$	X
4. A_1A_2	A_2	$0.5A_1A_2$	$0.5A_2A_2$	W_4	$0.5(u_{12} + u_{22})$	$0.5(u_{12} + u_{22})e_1$	X
5. A_2A_2	A_1	A_1A_2		W_5	u_{12}	u_{12}	1.0
6. A_2A_2	A_2		A_2A_2	W_6	u_{22}	u_{22}	V

$$\overline{W} D' = (W_1 D + 0.5\, W_3 H) p_m \qquad (1)$$

$$\overline{W} H' = (0.5\, W_3 H + W_2 R) p_m + (W_2 D + 0.5\, W_4 H) q_m \qquad (2)$$

$$\overline{W} R' = (0.5\, W_4 H) + W_6 R) q_m \qquad (3)$$

the mean fitness of the population is

$$\overline{W} = (W_1 D + W_3 H + W_2 R) p_m + (W_2 D + W_4 H + W_6 R) q_m \qquad (4)$$

The frequency of the A_1 allele in females is therefore

$$\overline{W} p'_f = (W_1 D + 0.75\, W_3 H + W_2 D) p_m + (0.5\, W_2 D + W_4 H) q_m \qquad (5)$$

and in males,

$$\overline{W} p'_m = [W_1 D + (0.5 + 0.25\psi) W_3 H + 0.5\, \psi\, W_2 H R] p_m$$

$$+ [(1 - 0.5\, \psi)\, W_2 D + (0.5 + 0.25\psi)\, W_4 H] q_m \qquad (6)$$

Thus female and male gene frequencies will generally be different as is common in sex-linked or haplodiploid systems. Therefore the first step in solving the recurrence equations (1), (2) and (3) is to eliminate this gene frequency difference by making the assumption that all males are worker-produced, hence $\psi = 1.0$. If this is done then equation (6) reduces to equation (5) so $p_m = p_f = p^*$ (say), and only a single gene frequency need be considered. I stress that this assumption, although biologically possible and true in a few species (e.g. Plateaux 1981), is made only for mathematical convenience so that the equations become analytically tractable. In fact the results only differ quantitatively when $\psi < 1.0$ (Owen 1986). An alternative approach is to use the ploidy-weighted average of the male and female gene frequencies (Abugov 1983). Another complication arises because with family or colony-level selection, unlike individual selection, genotype frequencies are not generally in Hardy-Weinberg proportions before selection. So the second step required to solve the equations is to introduce the quantity δ which measures the deviation of the genotypic frequencies from those expected. Therefore $D = p^{*2} + \delta$, $H = 2(p^* q^* - \delta)$, and

$R = q^{*2} + \delta$. The recurrence equations can now be solved (Owen 1986) to yield an equation for the equilibrium value (p) of p^*,

$$f(p) = [(2W_1 - W_2 - W_3)p^2 + 3(W_3 - W_4)pq + (W_2 + W_4 - 2W_6)q^2]pq + [(W_3 - W_2)p^2 + (2W_1 - 3W_3 + 3W_4 - 2W_6)pq + (W_2 - W_4)q^2]\delta = 0 \qquad (7)$$

with

$$\delta = (-B - \sqrt{B^2 - 4AC})/2A \qquad (8)$$

and,

$$A = (4W_3 - 2W_1 - 2W_2)p + (4W_4 - 2W_2 - 2W_6)q$$
$$B = -2W_1p^3 + (2W_1 - 8W_3)p^2q + (2W_6 - 8W_4)pq^2 - 2W_6q^3 + (W_3 - W_2)p + (W_4 - W_2)q$$
$$C = [2W_1p^3 + 2(W_2 + 2W_3)p^2q + 2(2W_4 + W_2)pq^2 + - (W_2 + W_3)p - (W_4 + W_2)q]pq.$$

Polymorphic equilibria

The equilibrium values of p are given by the roots of equation (7), which although rather complex, is actually only a cubic polynomial; hence a maximum of three polymorphic equilibria can exist. The number of realizable equilibria depends on the colony fitness model and the intensity of selection. Consider first non-additive fitnesses -- both the non-additive model and the special case of Crozier and Consul. When selection is strong, up to three equilibria can occur since equation (7) remains cubic (TABLE 3). However, when selection is weak the adult genotypic frequencies can be approximated by their Hardy-Weinberg proportions (Michod 1982, Abugov 1983); it follows therefore that the deviation from Hardy-Weinberg proportions, δ, can be taken to be approximately equal to zero. This means that the second term in equation (7), multiplied by δ, drops out and the equation reduces to a quadratic having at most two real roots.

The outcome is quite different with additive fitness values. Substitution of these into equation (7) gives,

$$(3pq + \delta) [(u_{12} - u_{11})p - (u_{12} - u_{22})q] = 0 \qquad (9)$$

TABLE 3. Some examples of polymorphic equilibria maintained by colony-level selection (from Owen 1986).

	u_{11}	u_{12}	u_{22}	e_1	e_2	$\hat{p}$	Stability	$\hat{\delta}$
1.	0.8	1.0	0.7	1.2	0.6	0.7940	Stable	-0.0169
2.	0.7	0.8	1.0	0.5	0.9	0.8662	Unstable	-
3.	0.8	0.6	0.5	1.0	2.0	0.4090	Stable	-0.0040
						0.7261	Unstable	-
4.	0.4	0.5	0.7	2.5	0.9	0.2221	Unstable	-
						0.6673	Stable	-0.0156
5.	0.5	1.0	0.5	2/15	2/15	0.1292	Stable	-0.0125
						0.5000	Unstable	-
						0.8708	Stable	-0.0125

which has only one root hence only a single nontrivial equilibrium can occur, regardless of selection intensity, at frequency

$$p = (u_{12} - u_{22})/(2u_{12} - u_{11} - u_{22}). \qquad (10)$$

The condition for stability is that $u_{12} > u_{11}, u_{22}$, which holds for all values of ψ (Owen 1986). This condition also implies that $u_{12} > 0.5 (u_{11} + u_{12})$, $0.5 (u_{12} + u_{22})$, i.e. the fitness of the heterogenotypic colony cannot exceed that of colonies composed of heterozygous workers. Hence with additive selection what could be described as worker heterosis is required for a stable polymorphic equilibrium. The situation is, of course, different with non-additive selection because in this case colony heterosis determines the existence and stability of equilibria (TABLE 3). It is the strength of the interaction term, e_i, that is important; meaning that stable equilibria can occur without worker heterosis.

Population mean fitness

The mean population fitness when $\psi = 1.0$ from equation (4) becomes

$$W = (W_1D + W_3H + W_2R)p^* + (W_2D + W_4H + W_6R)q^* \quad (11)$$

which is neither locally nor globally maximized at stable equilibrium with general or nonadditive selection (Owen 1986). This was proven for the autosomal case by Pollak (1978) and some specific examples for the haplodiploid case are given by Owen (1986). However, this result pertains only to the case of relatively strong selection. With weak selection when genotype frequencies can be approximated by Hardy-Weinberg proportions ($\delta = 0$) stable equilibria occur at peaks of mean fitness or fertility (Abugov 1983). Abugov (1983) has shown that with weak selection and general colony fitness (TABLE 1) the approximate change in allele frequency per generation is

$$\Delta p = \frac{pq\ \delta\overline{W}}{3W\ \delta p} \quad (12)$$

which defines a classical Wrightian adaptive topography. Hence "evolution will favor ... genes associated with increased family fertility ... and it will disfavor those associated with decreased family fertility" (Abugov 1983). It must be stressed that this is true only with weak selection for non-additive fitnesses. The situation differs with additive fitnesses. The mean population fitness (equation 11) reduces to

$$W = u_{11}p^{*2} + 2p^*q^*\ u_{12} + u_{22}\ q^{*2}, \quad (13)$$

which will be maximized at stable equilibrium regardless of the strength of selection. Hence the closeness of a given stable gene frequency equilibrium to a maximum of the mean population fitness depends on the strength of the interaction term e_i, and the strength of selection. The various possibilities are summarized in TABLE 4.

Multiple mating

Polyandry or multiple mating by females is widespread in social Hymenoptera (Page 1986), moreover unequal sperm usage is also common (Crozier and Brückner 1981). Multiple mating has no effect on the outcome of individual selection but is important for colony-level selection. Multiple mating results in a greater diversity of worker genotypes within colonies (Crozier and Page 1985) leading to an increase in the total number of colony phenotypes. For example, with double mating by queens and unequal sperm usage, there are 12 different mating-types and a minimum of 11 different colony fitnesses (TABLE 5). The effect of multiple mating is to increase the degree of the polynomial equilibrium equation for p (equation 7) with general nonadditive fitnesses. This means that the number of polymorphic equilibria possible is increased over single mating. Thus with double mating a maximum of four equilibria can occur with strong selection, and three with weak selection (Owen 1986). Again, however, with additive fitness only a single equilibrium is possible regardless of selection intensity.

Worker-produced males

The results so far derived have been based on the assumption that all males are produced by workers. This assumption was necessary for mathematical convenience to make the equations analytically tractable. Having all males produced by the queen or having a mixture of queen and worker-produced males causes a shift in the equilibrium frequencies. Also the male and female equilibrium frequencies are now, of course, different (Owen 1986). Moreover the conditions for stable polymorphic equilibria depend on the value of ψ. Thus the degree of male production by workers determines the possibility of the occurrence of stable polymorphic equilibria for a given set of colony fitness values. This can be clearly illustrated and quantified with the special case of colony viabilities of Crozier and Consul (1976). In this model the condition for at least one stable polymorphic equilibrium, for any value of ψ, is (Owen 1986),

$$X > (2V^2 - \psi V)/(1 + V - \psi)$$

FIGURE 1 shows the regions of fitness space in which stable and

TABLE 4. Relationship between mean population fitness at gene frequency equilibrium and strength of selection and interaction term e_i.

		$\lvert e_i \rvert$		
		>>1	~1	=1
Selection	Strong $\delta \neq 0$	Not Maximized	Approximately Maximized	Maximized
	Weak $\delta = 0$	Approximately Maximized	Approximately Maximized	Maximized

TABLE 5. Colony types in social Hymenoptera when queens mate twice and use sperm in proportion x_1 from the first male, and x_2 from the second male.

Mating type			Worker genotypes in colony			Colony
Queen	x	Male	A_1A_1	A_1A_2	A_2A_2	fitness
A_1A_1		A_1,A_1	1			W_1
A_1A_1		A_1,A_2	x_1	x_2		W_2
A_1A_1		A_2,A_1	x_2	x_1		W_3
A_1A_1		A_2,A_2		1		W_4
A_1A_2		A_1,A_1	0.5	0.5		W_5
A_1A_2		A_1,A_2	$0.5x_1$	0.5	$0.5x_2$	W_6
A_1A_2		A_2,A_1	$0.5x_2$	0.5	$0.5x_1$	W_7
A_1A_2		A_2,A_2		0.5	0.5	W_8
A_2A_2		A_1,A_1		1		W_4
A_2A_2		A_1,A_2		x_1	x_2	W_9
A_2A_2		A_2,A_1		x_2	x_1	W_{10}
A_2A_2		A_2,A_2			1	W_{11}

unstable equilibria occur for different values of ψ. For X, V < 1 the effect of worker-produced males is to relax the conditions necessary for a stable polymorphism. The important point to appreciate is that with colony-level selection no selective differences between male genotypes are assumed, yet, worker-produced males affect the outcome of selection.

Conclusions

Characters expressed by workers, whether or not they are also expressed by queens and males, can be selected for at the level of the colony. The outcome of colony-level selection depends on how the colony fitnesses are determined. If colony fitnesses are just the arithmetic average of the contributions of the worker genotypes (additive model), then only a single polymorphic equilibrium can occur. However, if colony fitness is determined by nonadditive interactions between worker genotypes, then a maximum of three (2 stable, 1 unstable or *vice versa*) can exist. An important result is that with strong selection and nonadditive colony fitnesses the mean population fitness will not necessarily be maximized at equilibrium. It will be approximately maximized with weak selection.

Although the models assume no selective difference between male genotypes, worker-produced males alter the conditions for the existence of a polymorphic equilibrium, and shift the male and female equilibrium allele frequencies. Multiple mating by queens is important because it increases the number of polymorphic equilibria possible.

How common is colony-level selection on unitary characters and how easily is it detectable? Undoubtedly characters uniquely possessed by workers must have evolved through colony-level selection (Darwin 1860), on the other hand, there is no reason why any character also expressed by queens and males, cannot be selected for at the level of the colony. In both cases it may be difficult to detect natural selection. Many of the methods, with their associated problems, for the detection of individual selection discussed by Endler (1986) can be applied equally well to colony-level selection. A potential problem is that the effects of individual and colony-level selection could be confounded for traits expressed by the reproductives as well as the workers. One promising avenue of approach is to compare observed and expected numbers of colony types in a population (Crozier 1973). Suppose that heterogenotypic colonies (TABLE 2) have a higher survival

45

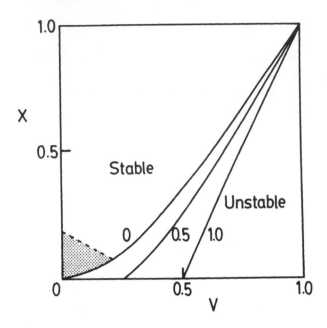

FIGURE 1. The range of colony fitness values that permit stable polymorphic equilibria for the special case considered by Crozier and Consul (1976). The curve $X = (2V^2 - \psi V)/(1+V-\psi)$ is shown for three values of ψ; 0 (the case given by Crozier and Consul), 0.5 and 1.0. Crozier and Consul found stable oscillations in the region indicated by the stippling, these are eliminated with worker-produced males.

rate than homogenotypic colonies, then their observed frequency would be greater than expected; moreover, the frequency of heterozygous foundress queens would be at an excess (Crozier 1973, Crozier and Consul 1976). Crozier (1973) found an example of this at a malate dehydrogenase isozyme locus in the ant *Aphaenogaster rudis*. At one of four locations in Georgia only heterogenotypic colonies were found, thus all queens were heterozygotes. The genotypic frequencies of the queens differed significantly from those at the other locations, from Hardy-Weinberg proportions, and from the worker genotypic frequencies. Similarly within colonies a deficiency or excess of one worker genotype could occur; for instance if one was more susceptible to temperature stress or had a greater tendency for foraging or defense (Crozier and Consul 1976). The relative fitness of a particular colony type of course could not be inferred from this observation, and caution must be used in the interpretation of worker genotype frequencies (Crozier and Consul 1976).

The theoretical consequences of colony-level selection on unitary characters have been fairly well explored, how it differs from individual selection has been revealed, and the importance of social structure (multiple mating and worker-produced males) has been shown. However finding examples of characters selected for at the colony level is more difficult, though as indicated here is possible. Characters that hold more potential for showing colony-level selection are the aggregate or group characters to be discussed in the next section.

AGGREGATE CHARACTERS

The characters discussed in the previous section are expressed genotypically and phenotypically by individual workers, and the colony phenotype depends simply on the proportion of each worker morph present. Some characters are only expressed by workers as a group. For instance, honey bee workers show nest thermoregulatory behavior only when members of a group and the temperature they maintain depends on group size (Free and Spencer-Booth 1958). Similarly alarm behavior in social insects, stimulated by the release of a pheromone, is by definition a group phenomenon. The reaction of individual workers depends on group size in honey bees (Southwick and Moritz 1983) and *Vespula* species (Moritz and Bürgin 1987). The allometric relationship between body dimensions of workers in a colony is an equally good example of a group character. The allometric relation between the dimensions of two body parts is given by

$$y = bx^k \qquad\qquad (15)$$

or alternatively,

$$\log y = \log b + k\log x$$

where y and x are measures of the body parts and k and b are the allometric constants. As workers vary in size and thus body dimensions, the allometric constants of interest can be calculated for each colony; clearly phenotypes can only be expressed by colonies not individuals. Allometric constants do differ among colonies of social insects. For instance Harder and Owen (unpublished) found significant differences for both intercept (b) and slope (k) of the regression of glossa length on wing length of workers between 15 colonies of the bumble bee *Bombus occidentalis* (FIGURE 2). It is of course essential to calculate biologically meaningful allometric constants. An allometric relationship may be calculated for any two body dimensions but this does not necessarily imply functional significance. In the case of bumble bees it does appear that body size and glossa length are important determinants of flower choice by workers (Morse 1978, Harder 1985).

In ants, caste polymorphism has long been recognized as a consequence of allometric growth (Huxley 1932). The head shows the greatest degree of allometric variation within the worker caste and this, coupled with the size frequency distribution, gives rise to caste polymorphism (Wilson 1953). In the simplest case of monophasic allometry, in which the regression line has a single slope (Wilson 1953), the values of the allometric coefficients are clearly important in determining caste proportions. A crucial question, posed by Crozier (1987) is "how easily can allometry curves be changed, (?)". The extent depends on the amount of genetic variation present for the allometric coefficients. Characters, such as these, derived from body dimensions are necessarily continuously varying, but so also are others like oxidative metabolism in honey bees (Southwick 1985) and alarm response (Moritz and Bürgin 1987), which are manifest by groups of workers. Methods and concepts of quantitative genetics are necessary for the analysis of continuous variation, but standard procedures used for estimating parameters of individually expressed characters must be modified for group characters. Moritz has (1986) made these modifications and the account given here is based on his paper. The method is analogous to standard sib-analysis (Falconer 1981, Oldroyd and Moran 1983) for the estimation of heritability. It depends on comparing groups of workers from the same colony to estimate within-colony variation for the group character, an intraclass correlation is then calculated from the between- and within-colony mean squares. Instead of measurements on individual workers each observation is a measurement made on a group of workers. Thus Moritz (1986) measured the change in O_2 consumption as a response to isopentyl-acetate by groups of workers taken from honey bee colonies.

The theory developed by Moritz (1986) is as follows; consider two groups (x and y) of workers, each of size n, from the

48

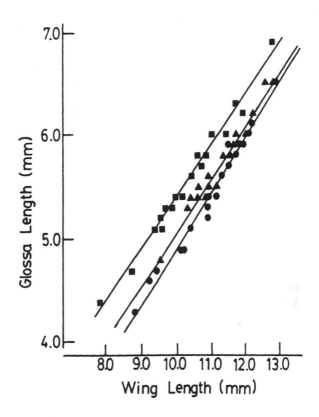

FIGURE 2. Regression of glossa length on wing length for 20 workers in each of three colonies of the bumble bee *Bombus occidentalis*; these differ significantly in intercept.

same colony. The average or additive relationship between the groups can be defined as,

$$r^* = (\Phi + \Phi')/2 \qquad (17)$$

where

Φ = probability that the paternal alleles are identical by descent in both groups, and

Φ' = probability that the maternal alleles are identical by descent in both groups.

Both probabilities depend on group size n, and are derived from the binomial distribution for the maternal alleles and the multinomial distribution for the parental alleles if queens have mated more than twice (Moritz 1986). [Note the close parallel to the definition of relatedness between individuals (Malécot 1969) $r = (\phi + \phi')/2$, where ϕ = probability that the paternal allele in one worker is identical to that in a sibling worker, and that ϕ' = a probability that the maternal allele in one worker is identical to that in a sibling worker (Oldroyd and Moran 1983)]. As Moritz (1986) notes the genetic difference between any two groups increases as group size decreases.

A one-way analysis of variance based on a linear model is used to estimate heritability. The phenotype of a group of workers is assumed to be determined by the sum of the individual phenotypes and the interactions between them (Moritz 1986). The total phenotypic variance (S^2_P) can be decomposed as follows,

$$S^2_P = S^2_A + S^2_I + S^2_E \qquad (18)$$

in which

S^2_A = additive genetic variance

S^2_I = total non-additive genetic variance

$\quad = S^2_D$ (dominance variance) + S^2_{GI} (genotypic interaction variance)

S^2_E = environmental variance.

These components of variance (S^2) are analogous to those for individual characters (V). The covariance between groups x and y,

$$\text{Cov } (x,y) = r^* S^2_A + \phi\phi' S^2_I. \qquad (19)$$

Heritability is estimated in a way similar to that for individual characters as

$$\hat{h} = t/r^* \qquad (20)$$

in which the intraclass correlation $t = \text{Cov } (x,y)/S^2_P$.

Hence, $\hat{h}^2 = S^2_A/S^2_P + \phi\phi'S^2_I/S^2_P.$ \qquad (21)

This heritability estimate derived by Moritz (1986) is analogous to that for individual characters based on sib analysis (Oldroyd and Moran 1983). Similarly the estimate of heritability in the narrow sense (S^2_A/S^2_P) may be biased upward by nonadditive genetic and dominance effects; this bias is given by the second term in equation (21.). The intraclass correlation (t) is calculated from the mean square estimates from a one-way analysis of variance (TABLE 6) of the data, r^* calculated from equation (17) for group size n, and then heritability from equation (20).

Moritz (1986) estimated the heritability of alarm behavior in the honey bee *Apis mellifera intermissa* from North Africa. He tested 15 groups (N), each of 40 workers (n), from 10 colonies (c). The intraclass correlation t ($\pm$ S.D.) = 0.84 $\pm$ 0.13, and the average relationship r^*, between the groups of 40 workers was calculated to be 0.875, as the *A. m. intermissa* queens were naturally mated with an unknown number of drones. Hence heritability

$\hat{h}^2$ = 0.84/0.875 = 0.96 $\pm$ 0.16 ($\pm$S.E.), with a potential bias of 48%.

This quantitative genetic approach offers great promise. It deals with characters expressed by workers as groups, that, therefore, represent properties of the colony as a whole. The characters encompassed range from morphological, to physiological and behavioral. Heritabilities of traits with clear ecological significance can be determined. Selection is expected to act on quantitative traits to change or maintain the mean and variance in the population. Also the response to selection can be predicted using standard quantitative genetic approaches (Falconer 1981).

CHARACTERS OF REPRODUCTIVES

In some instances a character expressed only by the reproductives (queens and males), and not by the workers, may be selected for at the level of the colony. For this to occur it is necessary that the fitness of the colony depends on the characters of the reproductives. Two examples will be considered the first being body size of reproductives.

Social insect colonies can be assumed to have only a finite amount of biomass or energy available to invest in reproductive offspring (Oster *et al.* 1977). The problems are allocation of this energy not only between, but also within, the sexes. The latter

51

TABLE 6. ANOVA table for estimation of variance components of group characters.

Source	d.f.	Expected M.S.
Between colonies	$c-1$	$k\sigma^2 + \sigma^2_w$
Within colonies	N_T-c-1	σ^2_w
TOTAL	N_T-1	

c = number of colonies
N_T = total number of groups samples
$k = [N_T - (\Sigma N_i^2/N_T)]/(c-1)$
N_i = number of groups sampled from i th colony.

concerns finding the optimal balance between size and number of offspring (Smith and Fretwell 1974), because with a fixed amount of energy available, increasing the number of offspring will cause a corresponding decrease in their individual size and *vice versa*. Assuming that the viability (w_i) of the reproductive offspring varies with body size (or mass) as in FIGURE 3, then the optimal body size of offspring can be found that maximizes the reproductive success of the colony. Consider for simplicity offspring of just one sex (males or queens), then let n = number, E = total energy of biomass available, and x = body size. Assuming body size is correlated with body mass, then $x \propto E/n$. The reproductive success of the colony, and hence founding queen and male, is nw_i which is maximized at the point where the tangent through the origin touches the curve in FIGURE 3 (Smith and Fretwell 1974). The result is an intermediate body size ($\hat{x}$) of offspring. This is an optimal solution from the point of view of the colony but obviously not from the perspective of the individual offspring, because individual selection occurring through viability differences will favor the largest individuals. As size is determined at the colony stage, colony-level selection will inevitably prevail. Owen (1988) interpreted body size of bumble bee (*Bombus*) queens in this way. For the simple case considered here, where only reproductive offspring of one sex are considered, the interests of the workers and the founding queen and male are identical. However when both sexes are considered there is a potential conflict of interest between the founding queen and the workers over the investment ratio of reproductives (Trivers and Hare 1976, Oster *et al.* 1977). Moreover Bulmer and Taylor (1981) have shown that if workers can vary investment in individual young queens and thus vary their viability, then the workers will gain greater control over the investment ratio. The investment ratio is also likely to be sensitive to the particular shapes of the viability functions (FIGURE 3) of the reproductive offspring (Macnair 1978). The important point is that whatever equilibrium investment ratio evolves, body size of reproductives is determined by colony-level selection.

The same results as those from the graphical model are also predicted by a simple explicit genetical model, combining productivity differences between colonies with viability differences between offspring. This is an extension of the general colony-level selection or fertility model given in TABLE 2, and some results have been obtained by Abugov (1985). In this case reproductive offspring of genotype $A_i A_j$ have viability w_{ij} independent of colony of origin. Abugov (1985) has shown that for weak selection,

$$\Delta p = \frac{pq}{3W_I{}^2 W_3} \frac{d W_c W_I{}^2}{dp} + 0(M^2) \qquad (22)$$

where $W_I = \sqrt{W_f\ W_m}$ is the mean viability of the individual reproductives in the population, and W_c is the mean fitness (productivity of the colonies in the population; equation 4). Hence maximization of $W^2{}_I W_c$ is predicted, which is interesting and important since it implies that selection gives twice the weight to increases in viability than to increases in fertility or colony productivity (Abugov 1985). Also Abugov (1985) points out that if viability of only one sex is affected by the locus in question then either $W_f W_c$ or $W_m W_c$ is maximized and viability and fertility receive equal weight. It must be remembered, as was shown earlier, that maximization will not necessarily occur with strong selection. The body size example can be conceptualized in terms of the single gene locus model. If only young queens are considered and if body size is determined by genotype, each has a different viability as shown in FIGURE 3. Because of the assumed inverse relationship between body size of offspring and colony

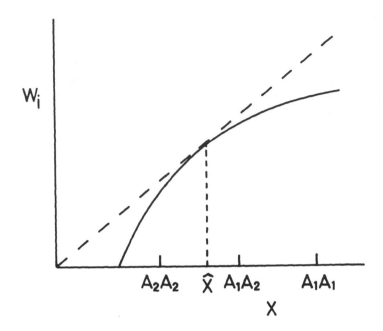

FIGURE 3. Fitness (w_i) of an individual reproductive offspring (queen or male) as a function of body size x, which may be determined by genotype (A_1A_1, A_1A_2 or A_2A_2). The optimal size $\hat{x}$ maximizes the reproductive success of the parental colony.

productivity, colonies yielding all A_1A_1 progeny (from A_1A_1 x A_1 matings) will be the least productive while those yielding all A_2A_2 offspring from (A_2A_2 x A_2 matings) will be the most productive. The other colony types will show various intermediate levels of productivity. Intuitively one would expect that, given appropriate values of productivities and viabilities, a stable polymorphic equilibrium could exist that maintains an intermediate body size.

Body size is an interesting example because it may be either controlled by the behavior of the workers, through regulation of larval feeding rates, or genotypes may have different intrinsic growth rates. In both cases, the reproductive success of the founding queen and male depend on the productivity of the colony and the viability of the reproductive offspring during their solitary phase.

The second example is multiple mating by queens. Crozier and Page (1985) evaluated some hypotheses for the evolution of polyandry in the social Hymenoptera. One hypothesis is that polyandry is favored because genetic variation within colonies is increased. Crozier and Page (1985) considered that a greater mix of worker genotypes is beneficial if environmental variation is large. This is equivalent to saying the more heterogenotypic the colony, the higher its fitness (compare TABLES 2 and 5). In this way, mating behavior of queens and males could be selected at the level of the colony.

SUMMARY

In this paper, I have attempted to show that many different types of characters of social insects can be analyzed under the rubric of colony-level selection. Moreover, the term colony-level selection, as used here, is given a specific and restrictive definition in terms of reproductive success of the founding queen and male. Therefore, colony-level selection can be viewed as mating-type or fertility selection. This is because characters expressed by workers or reproductive offspring are just a secondary expression of the genotypes of the founding queen and male. This is equally true for both unitary and aggregate characters of workers, although different methods are appropriate for their analysis.

The population genetic models for unitary characters reveal how the dynamics of colony-level and individual selection differ, and show the importance of social structure (multiple mating, worker-produced males). However it is another matter to actually demonstrate that a trait is or has been selected for at the level of the colony. In this regard the aggregate or group characters hold more potential. This is because a wide range of morphological, physiological and behavioral characters can be analyzed and the possibility of measuring fitness differences between colonies opens up. This is exciting because these aggregate characters represent the result of interactions between workers and are not expressed by individual workers themselves. Social insects exhibit some very complex feedback loops within colonies, including indications of colony "memory" which lasts longer than the lifespan of individuals (Gordon 1987). Lumsden (1982) has modelled the organization of colonies in terms of an ergonomic interaction matrix which controls the regulation of caste production. Wilson (1985) suggests that this interaction matrix is itself a character which is shaped by colony-level selection.

Acknowledgements

My research has been supported by the Natural Sciences and Engineering Research Council of Canada. I thank Laurence Packer for his comments on the manuscript, and Lawrence Harder for providing the regressions in FIGURE 2.

LITERATURE CITED

Abugov, R. 1983. Genetics of Darwinian fitness I. Fertility selection. *Am. Nat.* 121:880-886.

Abugov, R. 1985. Genetics of Darwinian fitness II. Pleiotropic selection on fertility and survival. *J. theor. Biol.* 114:109-125.

Bodmer, W. F. 1965. Differential fertility in population genetics models. *Genetics* 51:411-424.

Bulmer, M. G. and P. D. Taylor. 1981. Worker-queen conflict and sex ratio theory in social Hymenoptera. *Heredity* 47:197-207.

Butler, G. J., H. I. Freeman and P. Waltman. 1982. Global dynamics of a selection model for the growth of a population with genotypic fertility differences. *J. Math. Biol.* 14:25-35.

Cockerham, C., P. M. Burrows, S. S. Young and T. Prout. 1972. Frequency-dependent selection in randomly mating populations. *Am. Nat.* 106:493-515.

Crozier, R. H. 1973. Apparent differential selection at an isozyme locus between queens and workers of the ant *Aphaenogaster rudis. Genetics* 73:313-318.

Crozier, R.H. 1987. Towards a sociogenetics of social insects. In *Chemistry and biology of social insects.* J. Eder and H. Rembold, eds. Verlag J. Peperny:München pp. 325-328.

Crozier, R. H. and D. Brückner. 1981. Sperm clumping and the population genetics of Hymenoptera. *Am. Nat.* 117:561-563.

Crozier, R. H. and P. C. Consul. 1976. Conditions for genetic polymorphism in social Hymenoptera under selection at the colony level. *Theor. Pop. Biol.* 10:1-9.

Crozier, R. H. and R. E. Page. 1985. On being the right size: male contributions and multiple mating in social Hymenoptera. *Behav. Ecol. Sociobiol.* 18:105-115.

Darwin, C. 1860. *On the origin of species.* 2nd ed. John Murray:London.

Endler, J. A. 1986. *Natural selection in the wild.* Princeton University Press:Princeton.

Falconer, D. S. 1981. *Introduction to quantitative genetics.* 2nd. ed.

Longmans:London.

Feldman, M. W. and U. Liberman. 1985. A symmetric two locus fertility model. *Genetics* 109:229-253.

Feldman, M. W., F. B. Christiansen and U. Liberman. 1983. On some models of fertility selection. *Genetics* 105:1003-1010.

Free, J. B. and H. Y. Spencer-Booth. 1958. Observations on the temperature regulation and food consumption of honey-bee (*Apis mellifera*). *J. Exp. Biol.* 35:930-937.

Gordon, D. M. 1987. Group-level dynamics in harvester ants: young colonies and the role of patrolling. *Anim. Behav.* 35:833-843.

Hadeler, K. P. and U. Liberman. 1975. Selection models with fertility differences. *J. Math. Biol.* 2:19-32.

Harder, L. D. 1985. Morphology as a predictor of flower choice by bumble bees. *Ecology* 66:198-210.

Howard, D. R. and W. R. Tschinkel. 1976. Aspects of necrophoric behaviour in the red imported fire ant *Solenopsis invicta*. *Behaviour* 56:157-180.

Huxley, J. 1932. *Problems of relative growth.* Methuen:London.

Liberman, J. and M. W. Feldman. 1985. A symmetric two locus model with viability and fertility selection. *J. Math. Biol.* 22:31-60.

Lumsden, C. J. 1982. The social regulation of physical caste: the superorganism revived. *J. theor. Biol.* 95:749-781.

Macnair, M. R. 1978. An ESS for the sex ration in animals, with particular reference to the social Hymenoptera. *J. theor. Biol.* 70:449-459.

Malécot, G. 1969. *The mathematics of heredity.* W.H. Freeman:San Francisco.

Michod, R. E. 1982. The theory of kin selection. *Ann. Rev. Ecol. Syst.* 13:23-55.

Moritz, R. F. A. 1986. Estimating the genetic variance of group characters: social behaviour of honeybees (*Apis mellifera* L.). *Theor. Appl. Genet.* 72:513-517.

Moritz, R. F. A. and H. Bürgin. 1987. Group response to alarm pheromones in social wasps and the honeybee. *Ethology* 76:15-26.

Morse, D. H. 1978. Size-related foraging differences of bumble bee workers. *Ecol. Entomol.* 3:189-192.

Nagylaki, T. 1979. Selection in dioecious populations. *Ann. Hum. Genet.* 43:143-150.

Nagylaki, T. 1987. Evolution under fertility and viability selection. *Genetics* 115:367-375.

Oldroyd, B. and C. Moran. 1983. Heritability of worker characters in the honey bee (*Apis mellifera*). *Aust. J. Biol. Sci.* 36:323-332.

Oster, G., I. Eshel and D. Cohen. 1977. Worker-queen conflict and the evolution of social insects. *Theor. Pop. Biol.* 12:49-85.

Owen, R. E. 1986. Colony-level selection in the social insects: single locus additive and non-additive models. *Theor. Pop. Biol.* 29:198-234.

Owen, R. E. 1988. Body size variation and optimal body size of bumble bee queens (Hymenoptera:Apidae). *Can. Entomol.* 120:19-27.

Page, R. E. 1986. Sperm utilization in social insects. *Ann. Rev. Entomol.* 31:297-320.

Penrose, L. S. 1949. The meaning of "fitness" in human populations. *Ann. Eugenics.* 14:301-304.

Plateaux, L. 1981. The *Pallens* morph of the ant *Leptothorax nylanderi*: description, formal genetics, and study of populations. In *Biosystematics of social insects,* P.E. Howse and J.-L. Clement, eds. Academic Press:New York pp. 63-74.

Pollak, E. 1978. With selection for fecundity the mean fitness does not necessarily increase. *Genetics* 90:383-389.

Rothenbuhler, W. C. 1964. Behaviour genetics of nest cleaning in honey-bees. I. Responses of four inbred lines to disease-killed brood. *Anim. Behav.* 12:578-583.

Roux, C. Z. 1977. Fecundity differences between mating pairs for a single autosomal locus, sex differences in viability and nonoverlapping generations. *Theor. Pop. Biol.* 12:1-9.

Seeley, T. D. 1983. Division of labor between scouts and recruits in honeybee foraging. *Behav. Ecol. Sociobiol.* 12:253-259.

Smith, C. C. and S. D. Fretwell. 1974. The optimal balance between size and number of offspring. *Am. Nat.* 108:499-506.

Sober, E. 1984. *The nature of selection.* MIT Press:Cambridge, Mass.

Southwick, E. E. 1985. Allometric relations, metabolism and heat conductance in clusters of honeybees at cool temperatures. *J. Comp. Physiol. B.* 156:143-149.

Southwick, E. E. and R. F. A. Moritz. 1985. Metabolic response to alarm pheromone in honeybees. *J. Insect Physiol.* 31:389-392.

Starr, C.K. 1979. Origin and evolution of insect sociality: a review of modern theory. In *Social insects, Volume 1,* H. R. Hermann, ed. Academic Press:New York pp. 35-79.

Templeton, A. R. 1979. A frequency dependent model of brood selection. *Am. Nat.* 114:515-524.

Trivers, R. L. and H. Hare. 1976. Haplodiploidy and the evolution of the social insects. *Science* 191:249-263.

Visscher, P. K. 1983. The honeybee way of death: necrophoric behaviour in *Apis mellifera* colonies. *Anim. Behav.* 31:1070-1076.

Wade, M. J. 1978. A critical review of the models of group selection. *Quart. Rev. Biol.* 53:101-114.

Wade, M. J. 1980. Kin selection: its components. *Science* 210:665-667.

West-Eberhard, M. J. 1975. The evolution of social behaviour by kin selection. *Quart. Rev. Biol.* 50:1-33.

West-Eberhard, M.J. 1979. Intragroup selection and the evolution of insect societies. In *Natural selection and social behaviour: recent research and theory*, R.D. Alexander and D.W. Tinkle, eds. Chiron Press:New York. pp. 3-17.

Wilson, E. O. 1953. The origin and evolution of polymorphism in ants. *Quart. Rev. Biol.* 28:136-156.

Wilson, E. O. 1971. *The insect societies.* Belknap Press of Harvard University Press:Cambridge, Mass.

Wilson, E. O. 1985. The principles of caste evolution. In *Experimental behavioural ecology and sociobiology*, B. Hölldobler and M. Lindauer, eds. Sinauer Associates, Inc.:Sunderland pp. 307-324.

Wilson, E. O., N. E. Durlach and L. M. Roth. 1958. Chemical releasers of necrophoric behaviour in ants. *Psyche* 65:108-114.

CHAPTER 4

GENETIC BASIS FOR DIVISION OF LABOR IN AN INSECT SOCIETY

Gene E. Robinson and Robert E. Page, Jr.

Some of the most conspicuous traits of social insects, such as nest construction and maintenance, defense, and foraging, are a consequence of the collective endeavors of a great number of individuals. The processes that integrate worker behavior into coordinated colony patterns are poorly understood. A colony can perform a variety of complex activities, sometimes simultaneously, in part due to a crucial feature of its organization: division of labor among workers. Insight can be gained into the mechanisms and evolution of colony behavior by identifying the factors that determine the activities of individual workers.

In many groups of social insects there are broad patterns of division of labor based on differences in worker morphology or age (e.g., ants: reviewed by Oster and Wilson 1978; bumble bees: Cameron, in press; honey bees: reviewed by Seeley 1985; Winston 1987; stingless bees: Sommeijer 1983; wasps: reviewed by Jeanne 1980; Jeanne *et al.*, 1988; Post *et al.*, 1988; termites: Jones 1980). Based on this body of work Wilson (1985) argued that the proximal determinants of worker behavior are primarily, if not exclusively, extrinsic.

There is, however, increasing evidence (reviewed by Oster and Wilson 1978; Jeanne *et al.* 1988) for differences in behavior among individual colony members that are not attributable to worker morphology or age. Colony members differ genetically due to recombination, polyandry, and polygyny, suggesting a genetic basis for interindividual behavioral variation (Blum 1977; Crozier and Consul 1976; Nowogrodzki 1984; Owen 1986). This suggestion has become more compelling in light of recent studies demonstrating that the genetic structure of an insect colony can influence many social interactions, including those that result in

Department of Entomology, The Ohio State University, Columbus, OH 43210 USA differential reproduction (Metcalf and Whitt 1977a, 1977b; Crozier *et al.* 1984), kin recognition (reviewed by Gamboa *et al.* 1986; Breed and Bennett 1987; Page and Breed 1987), and conflict over the investment sex ratio (Trivers and Hare 1976). Nevertheless, the possible effects of colony genetic structure on the division of labor have received relatively little attention (Crozier and Consul 1976; Owen 1986; Moritz and Southwick 1987).

A genetic basis for division of labor in the honey bee, *Apis mellifera* is supported by new studies (Calderone and Page 1988; Frumhoff and Baker 1988; Robinson and Page 1988). Here we demonstrate that genetic differences between related colony members affect the division of labor within honey bee colonies (see also Robinson and Page, 1988). We also discuss how genetic regulation of the task performance of individual workers may influence colonial patterns of behavior. The implications of these results for the evolution of colony organization are addressed in Chapter 2.

HONEY BEE COLONY ORGANIZATION

More is known about the social and genetic organization of honey bee colonies than any other species of eusocial insect. This knowledge, coupled with the ability to manipulate colony genetic structure with instrumental insemination of queens (Laidlaw 1977), provides the foundation for genetic studies of division of labor and is reviewed briefly.

Honey bee division of labor is elaborate. Individual workers exhibit age polyethism and pass through a sequence of behavioral phases or "temporal castes" (Oster and Wilson 1978) during their 4- to 7-week life, each characterized by the performance of one or more tasks (reviewed by Seeley 1985; Winston 1987). Younger honey bee workers labor in the nest while older individuals forage, a pattern thought to be nearly universal among eusocial insects (Wilson 1971).

Honey bee behavior also varies independently of age polyethism. The rate of behavioral ontogeny varies; some individuals develop precociously and begin foraging earlier in life, while others mature more slowly and spend a greater amount of time working in the nest (Sekiguchi and Sakagami 1966; Nowogrodzki 1983). Individuals of similar ages differ in the extent to which they perform certain tasks (reviewed by Seeley 1985; Winston 1987). Only a fraction of a colony's workers engage in "rare" tasks, such as guarding the nest entrance or removing corpses. Some workers perform these tasks repeatedly while most never engage in them (Moore *et al.* 1987; Visscher 1983). In contrast, other jobs such as brood care and nectar foraging are thought to involve virtually all workers at some point in their lives (Sekiguchi and Sakagami 1966).

Honey bee colonies are genetically complex due to polyandrous mating. Queen honey bees mate with from 7 to 17 different drones (reviewed by Page 1986) and use the sperm of at least several drones at any one time (Page and Metcalf 1982; Laidlaw and Page 1984; Moritz 1985). Colonies thus consist of numerous subfamilies (Laidlaw 1974).

EXPERIMENTAL COLONIES USED IN GENETIC STUDIES OF DIVISION OF LABOR

We hypothesized that workers within a subfamily (super sisters) are more similar in behavior than are workers belonging to different subfamilies. To test this hypothesis workers from five colonies composed of electrophoretically distinguishable subfamilies were observed performing specific tasks and collected. Analyses of *malate dehydrogenase* (*Mdh*) allozymes were used to determine whether members of some subfamilies were more likely to be sampled at certain tasks than were members of other subfamilies.

Each experimental colony was composed of a queen, instrumentally inseminated with the semen of three genetically distinct drones, and worker progeny belonging to the three resultant subfamilies. Semen from each drone trio was pooled, diluted, and homogenized to stabilize subfamily frequencies over time (Kaftanoglu and Peng 1980; Moritz 1983). Subfamily membership could be determined by polyacrylamide gel electrophoresis because each drone used for insemination possessed a different *Mdh* allozyme, "slow" (S), "medium" (M), or "fast" (F) (Contel *et al.* 1977).

Techniques for studying the *Mdh* locus in honey bees are well established (Contel *et al.* 1977; Nunamaker and Wilson 1980; Del Lama *et al.* 1985). *Mdh* alleles undergo Mendelian segregation and the electrophoretic mobilities of the allozymes they code for are known for all honey bee life stages. *Mdh* allozymes provide consistent, unambiguous phenotypes, enabling scoring on gels with a high degree of certainty (FIGURE 1; see also Nunamaker and Wilson 1980; Page and Metcalf 1982). In addition, allozymes are

63

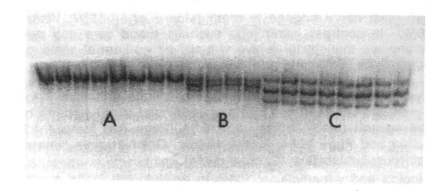

FIGURE 1. Allozyme markers used to identify subfamilies in honey bee colonies. Photograph of polyacrylamide gel stained for *malate dehydrogenase* (*Mdh*) revealing three easily distinguishable worker honey bee allozyme phenotypes, named on the basis of electrophoretic mobility. A: "slow-slow" (S); B: "slow-medium" (M); C: "slow-fast" (F). Samples are from a colony derived from a queen instrumentally inseminated with the semen of three drones, each bearing a different *Mdh* allozyme.

preferred as markers to visible mutants because they allow behavioral observations to be conducted blindly. They are reliable genetic markers and useful tools in behavioral studies.

TASKS STUDIED

Analyses of four tasks are presented: guarding the nest entrance, removing corpses from the nest, foraging for pollen, and foraging for nectar. Guarding and "undertaking", two rare tasks, are performed at similar ages, just prior to the onset of foraging (Sakagami 1953; Visscher 1983; Moore *et al.* 1987). Although only a few percent of a colony's workers ever engage in these tasks, both are important to colony survival (Visscher 1983; Moore *et al.* 1987). Guarding is a component of colony defense, while corpse removal promotes nest hygiene. Pollen foraging is not as rare as guarding or undertaking, but apparently is a more restricted activity than nectar collection. Extensive observations of individually marked workers indicate that a sizeable fraction of a colony's foraging force (28-33%) never collect pollen (Ribbands 1952; Sekiguchi and Sakagami 1966), while virtually all workers gather nectar (Sekiguchi and Sakagami 1966).

Workers engaged in these four tasks were unambiguously identified according to the following criteria. **Guards** Bees observed at the hive entrance with open mandibles, prothoracic legs off the substrate, antennae directed forward, wings elevated away from their bodies, quickly approaching and antennating bees landing at the nest entrance (Moore *et al.* 1987). **Undertakers** Workers observed leaving the hive carrying a corpse. To stimulate the performance of this task, 100 freshly frozen worker bees were introduced prior to the start of observations (Gary 1960; Visscher 1983). **Pollen Foragers** Bees returning to the hive carrying loads of pollen in their corbiculae were identified as pollen foragers. **Nectar foragers** were initially identified by their distended abdomens when returning to the hive. This initial determination was verified by measuring the volume of the foregut load; only bees carrying more than 10 μl were designated as foragers (Gary and Lorenzen 1976). Analysis of expressed crop contents using a refractometer revealed whether they were collecting water or nectar (greater than 5% sugar; Lindauer 1955).

Workers (n = 40) were collected while performing each task. Each colony was sampled twice, at 10- to 14-day intervals, for all tasks. "Control" samples, to determine the proportion of adults from each subfamily in the colony, were collected randomly with respect to behavior from frames of honey in the nest periphery, where bees of similar ages to guards and undertakers have been shown to be located (Seeley 1982). Foragers were collected about two weeks later than control samples, so control bees were also similar in age to the bees in the forager samples and most likely would have been part of their colony's foraging force had they not been removed earlier. Control samples provided an appropriate reference population for guards, undertakers, and foragers in the event of short-term fluctuations in subfamily frequencies due to non-random patterns of sperm use by queens or to differential worker mortality.

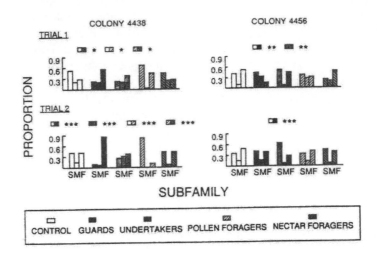

FIGURE 2. Genotypic composition of samples of task performers and control workers from two honey bee colonies composed of electrophoretically distinguishable subfamilies. Statistical analyses: each small rectangle above a graph denotes a comparison between two worker groups (e.g., white and black rectangles = controls vs. guards); * = $p < 0.05$; ** $p < 0.01$; *** $p < 0.001$, based on G-tests of actual frequencies. N = 36-40 workers per sample. Intertrial differences: for guards, Colony 4438: $p < 0.02$; Colony 4456: $p < 0.05$; for pollen foragers, Colony 4456: $p < 0.05$. Modified from Robinson and Page (unpublished data).

SUBFAMILY DIFFERENCES IN TASK PERFORMANCE

Results from two representative colonies are shown in FIGURE 2. Analyses of all five colonies (guards and undertakers: Robinson and Page 1988; foragers: Robinson and Page, unpublished data) reveal significant differences in the subfamily composition of samples of guards and control bees a total of 9 out of 10 times (5 out of 5 colonies, two trials each); 4 out of 10 times in samples of undertakers and control bees (3 out of 5 colonies); 5 out of 10 times in samples of guards and undertakers (5 out of 5 colonies); 6 out of 10 times in samples of pollen foragers and control bees (4 out of 5 colonies); 3 out of 10 times in samples of nectar foragers and control bees (2 out of 5 colonies); and 4 out of 10 times in samples of pollen foragers and nectar foragers (4 out of 5 colonies). These differences occurred too frequently to be explained by sampling error alone (nectar foragers vs. controls, $p < 0.05$; undertakers vs. controls, $p < 0.01$; all others, $p < 0.001$; binomial tests, based on the assumption of independence for the two trials from each colony).

The genotypic composition of control samples did not vary for any colony between trials 1 and 2. This measure of temporal subfamily homogeneity demonstrates that the relative frequencies of subfamilies within colonies were stable over the study period. Thus possible age differences between control bees and task performers, coupled with age polyethism, cannot account for the observed differences in subfamily representation.

Subfamilies differ markedly in their patterns of labor allocation (FIGURE 3). This is revealed by calculating the relative likelihood (RL) that a bee observed performing a particular task belongs to a given subfamily, according to the following formula.

$$RL_i = \frac{L_i}{L_T}$$

Where:

$L_i =$ <u>proportion of Subfamily i or j in task performer sample</u>

proportion of Subfamily i or j in control sample

$L_T =$ sum of L_i over all subfamilies within a colony

This procedure adjusts for differences in subfamily size and allows direct comparisons between the performance of a colony's subfamilies. There are differences in the likelihood of guarding and undertaking, two tasks performed by workers approximately the same age. Some subfamilies exhibit a relatively high probability of performing one task but not the other, while others perform both with similar likelihood. A different pattern is apparent for the two foraging activities, both performed by relatively older bees. Subfamilies appear more or less equally likely to forage for nectar, but diverge in pollen foraging. An explanation for variation in the degree of subfamily differentiation based on task type is proposed below.

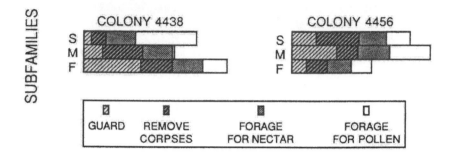

FIGURE 3. Subfamily differences in labor allocation. Within each horizontal bar, the width of a box represents the relative likelihood (RL) that a bee observed performing one of four tasks belongs to a given subfamily. For each task, RL values from all three subfamilies sum to one. Subfamilies with the highest cumulative RL values for all four tasks have the longest bars. See text for derivation of RL.

Intertrial differences in the genotypic composition of samples of task performers (FIGURE 2) suggest environmental effects on the probability of task performance. A strong genotypic component to the observed subfamily variation is demonstrated by consistent subfamily biases from trial to trial. This was quantified (FIGURE 4) by using analysis of variance (ANOVA) to calculate the proportion of the variance in RL due either to differences between subfamilies (genetic component) or between trials (environmental component and/or sampling error). There was relatively little change in subfamily patterns of task performance from trial to trial for guarding and undertaking, and less consistency for pollen and nectar foraging.

Further evidence for highly constant subfamily biases in task performance behavior was obtained by again sampling undertakers from colony 4456 four months after the first

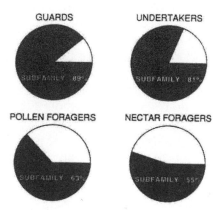

FIGURE 4. Consistent subfamily biases for task performance. Results of ANOVAs indicating the proportion of the variance in RL (relative likelihood that a bee observed performing a particular task belongs to a given subfamily) due to either differences between trials (white portions of circles) or repeated differences in task performance between a colony's subfamilies (black portions). Analyses were performed on angular transformations of RL values.

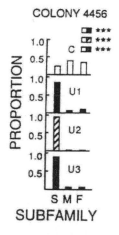

FIGURE 5. Persistence of subfamily differences in task performance: genotypic composition of three samples of undertakers (U1-3) and a sample of control bees (C) collected four months after the first samples from this colony were taken (FIGURE 3, Colony 4456). Explanation of statistical analyses in the legend for FIGURE 2. From Robinson and Page (1988). Reprinted by permission from Nature © 1988 Macmillan Magazines Ltd.

undertaker samples were collected. Bees from the S subfamily were consistently over-represented in the undertaker groups relative to the control sample (FIGURE 5), as in the earlier trials (FIGURE 2). The consistency of this effect, observed over the span of several worker lifetimes (despite differences in colony and environmental conditions associated with changing seasons), points to a strong genetic component for undertaking.

The colonies used in these studies consisted of three subfamilies, which is fewer than the number found in colonies with naturally mated queens. This raises the possibility that our results are an artifact of an unusually simple colony genetic structure. Recently, similar subfamily differences in task performance have been found with allozyme analyses of two colonies derived from naturally mated queens (N.W. Calderone, G.E. Robinson, and R.E. Page, unpublished data), validating our findings.

There are at least two possible explanations for our results. Genetically determined differences in subfamily spatial distribution due to location preferences within the nest may affect the probability of encountering task-related stimuli. However, super sisters were not clumped within honey bee nests (Frumhoff and Schneider 1987). We suggest instead that a colony's subfamilies differ in task performance because they have different distributions of behavioral response thresholds for task-eliciting stimuli. This hypothesis is consistent with previous reports of genotypic (Collins 1979) and hormonally mediated (Robinson 1987) differences in honey bee sensitivity to task-associated stimuli. Evidence for genotypic differences in response thresholds can also be inferred from studies on genetic regulation of honey bee hygienic behavior (Rothenbuhler 1964).

SUBFAMILY-BASED COLONY ORGANIZATION

Our results (see also Calderone and Page 1988; Frumhoff and Baker 1988; Robinson and Page 1988) demonstrate that there is a genetic basis for differences in task performance among individual colony members. Because colony behavior is a result of the behavior of its members, this discovery suggests that intracolonial genetic variation in task performance may underlie behavioral phenomena that are observed at the colonial level. Below we discuss the possible influence of intracolonial genetic variation in task performance on several behavioral traits of colonies. In shifting the focus from the individual worker to the whole colony we adopt a new view of the honey bee society: an ensemble of subfamilies, groups of workers with different distributions of genetically determined response thresholds to task-related stimuli. Implicit in our discussion is the assumption that there is also behavioral variation within subfamilies through recombination and maternally inherited genomes.

Behavioral Dominance

For any given behavioral trait, the colony's "phenotype" is determined by the behavior of its subfamilies. If each individual's phenotype is a consequence of additive effects of its genotype and environment, and each individual contributes additively to some colony phenotype, then additive effects of all subfamilies will define the colony phenotype. These assumptions may be most likely to hold for traits influenced by a large number of workers. For tasks performed by only a fraction of a colony's workers, interactions between individuals may result in non-additive effects on the colony phenotype, such as "behavioral dominance" (Craig 1980). Performance of a relatively rare task by individuals with the lowest response thresholds for that task, i.e., those most sensitive to the task-related stimuli (FIGURE 6a), may help maintain the stimulus level below the thresholds of less sensitive individuals, further diminishing the probability that less sensitive workers will perform that task (FIGURE 6b).

A worker genotype that results in an extremely low response threshold for a task may strongly influence the colony's performance of that task even if possessed by a minority of workers in a colony. This suggests that for some tasks the effects of behavioral dominance and negative feedback may be such that the variance in a colony's worker genotypes is an important determinant of the colony phenotype.

Negative feedback loops, already implicated as important regulators of colony activity (Wilson 1985; Seeley and Levien 1987), may exert especially strong effects on the genotypic structure of groups performing rare tasks because the activities of a small number of workers may satisfy colony needs. Our results suggest that genotypic differentiation is greater for rare tasks than for tasks performed by a large portion of colony members. Subfamily differences for guarding and undertaking, two rare tasks, were more consistent from trial to trial than were the subfamily differences observed for nectar and pollen foraging (FIGURE 4), which are more common tasks. Furthermore, pollen

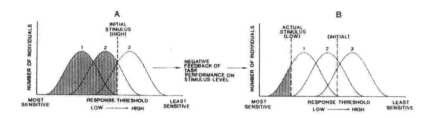

FIGURE 6. Effects of behavioral dominance and negative feedback loops of task regulation on the genotypic composition of a group of task performers in a hypothetical colony composed of three subfamilies (distributions 1-3). A) initial stimulus level associated with a particular task in a colony; B) actual stimulus level in the colony as a consequence of negative feedback. Shaded portions indicate the proportion of each subfamily's workers performing the task.

foragers showed greater genotypic differentiation relative to control samples than did nectar foragers (FIGURE 2), and as discussed earlier, pollen foraging apparently involves fewer workers than does nectar foraging. These post-hoc comparisons are obviously not conclusive, but the trends are in the predicted direction.

The Performance of Rare Tasks and Worker Inactivity

A common but unexplained characteristic of insect colonies is that worker activity levels are not uniform. Some workers exhibit relatively high activity levels while others appear relatively idle (e.g., Oster and Wilson 1978). Based on the assumption that a colony consists of a collection of distributions of workers with different mean response thresholds (e.g., subfamilies in polyandrous societies), individuals with a wide range of behavioral phenotypes may be present. A mechanistic explanation can thus be offered for one of the most enigmatic forms of division of labor in insect societies, the occurrence of workers that perform tasks most colony members do not, like guarding and undertaking. Individuals performing rare tasks have rare genotypes that result in extremely low response thresholds (FIGURE 7). This model also predicts the existence of inactive "reserve" workers (FIGURE 7), whose presence in honey bee colonies has previously been suggested (Sekiguchi and Sakagami 1966; Kolmes 1985; Fergusson and Winston, in press). Inactive workers may have response thresholds so high that they are usually insensitive to the task-related stimuli they encounter. Differences between subfamilies in the proportion of individuals with extreme genotypes may be a consequence of differences in either the response threshold mean or variance for each subfamily (FIGURE 8).

Social Homeostasis

The ability of colonies to respond to changes in environmental and social conditions by altering the ratio of individuals performing various tasks (Wilson 1983) may be in part a consequence of intracolonial genetic variation in worker behavior. For example, under "normal" conditions (FIGURE 9a), workers with the lowest response thresholds perform a task. If the need for this task increases due to changes in colony and/or environmental conditions, and there is a concomitant rise in the levels of associated stimuli (FIGURE 9b), then workers with relatively higher response thresholds perform the task. Through this mechanism moderate, transient, increases in the need for a particular task would cause differential recruitment among subfamilies, as more individuals shift from one job to another within the behavioral repertoire of a temporal caste.

A colony responds differently to severe, long-term changes in colony conditions, and this type of colony behavior may also be based on genotypic differences between workers. Changes in colony demography result in altered patterns of age polyethism in some worker honey bees (reviewed by Winston 1987) as a consequence of environmental modulation of worker bee juvenile hormone titers (Robinson *et al.* 1988; G.E. Robinson, C. Strambi, and A. Strambi, unpublished data). Workers from some subfamilies are more likely to display this form of behavioral plasticity than are workers from other subfamilies (G.E. Robinson and R.E. Page, unpublished data).

73

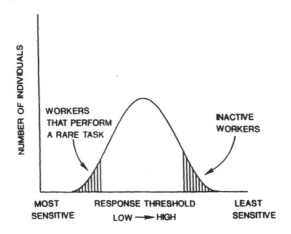

FIGURE 7. Hypothetical distribution of behavioral response thresholds for a subfamily. The behavior of bees that perform rare tasks or remain inactive may be a consequence of their possessing rare genotypes that result in extreme behavioral response thresholds.

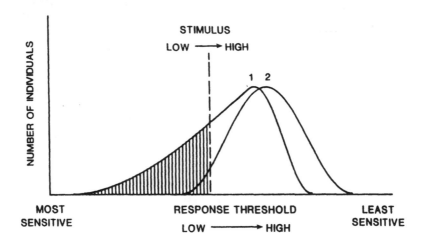

FIGURE 8. Hypothetical distributions of behavioral response thresholds for two subfamilies (distributions 1 and 2) showing differences in the proportion of individuals with extreme genotypes, due to differences in response threshold variance within each subfamily. Shaded portions indicate the proportion of each subfamily's workers performing the task.

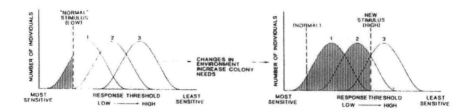

FIGURE 9. Effects of intracolonial variation in behavioral response thresholds on social homeostasis in a hypothetical colony composed of three subfamilies (distributions 1-3). A) "normal" conditions, with task performed by workers with lowest response thresholds; B) high stimulus conditions due to increased colony needs: workers with relatively higher response thresholds are now involved. Shaded portions indicate the proportion of each subfamily's workers performing the task.

If adaptive colony responses to environmental change are in part a consequence of intracolonial genotypic variation, then colonies with greater genetic heterogeneity may operate more smoothly in variable environments (Crozier and Consul 1976; Crozier and Page 1985), especially if a collection of specialized genotypes performs more efficiently under a range of environmental conditions than a single more generalized genotype. However, as discussed in Chapter 2, it is also possible that worker genetic variation exists for reasons independent of social homeostasis.

CONCLUSIONS

The discovery that genetic differences between workers influence the division of labor challenges the prevailing model of highly eusocial insect colonies as superorganisms in which individual members are equally likely to perform all necessary jobs. Worker honey bees exhibit impressive behavioral plasticity (reviewed by Winston 1987) but do not converge on some average level of task performance in a colony. As a consequence, differences between colonies are in part due to differences in the subfamily distribution of worker traits within colonies. Intracolonial differences in behavior thus underlie behavioral variation among and between populations and races of honey bees (reviewed by Rinderer and Collins 1986; see also Winston and Katz 1982; Calderone and Page 1988).

One of the more intriguing implications of a subfamily-oriented analysis of colony organization is that at least some behavioral traits of colonies, such as the occurrence of workers specializing on rare tasks and social homeostasis, may be based, in part, on genotypic differences between individual workers. Further elucidation of the genetic processes that underlie worker differentiation may lead to a better understanding of colony behavior. It remains to be seen whether colony genetic structure influences the division of labor in other insect societies.

Acknowledgements

We thank M.K. Fondrk for expert technical assistance during all phases of this study, J.A. Gianelos for help with electrophoretic analyses, N.W. Calderone, J.F. Downhower, T. Prout, T.D. Seeley, and P.W. Sherman for valuable discussions, and M.D. Breed, S.A. Cameron, and J.F. Downhower for reviewing the manuscript. Supported in part by an Ohio State University Postdoctoral Fellowship to GER, and NSF grants BNS-86056041 and BNS-8719283 to REP.

LITERATURE CITED

Blum, M. S. 1977. Ecological and social aspects of the individual behavior of social insects. *Proc. 8th Int. Congr. IUSSI, Wageningen, Holland.* pp. 54-59.

Breed, M. D. and B. Bennett. 1987. Kin recognition in highly eusocial insects. In *Kin recognition in animals*, D. J. C. Fletcher and C. D. Michener, eds. John Wiley & Sons Ltd.:New York pp. 243-285.

Calderone, N. W. and R. E. Page. 1988. Genotypic variability in age polyethism and task specialization in the honey bee, *Apis mellifera* (Hymenoptera: Apidae). *Behav. Ecol. and Sociobiol.* 22:17-25.

Cameron, S. A. Temporal patterns of division of labor among workers in the primitively eusocial bumblebee *Bombus griseocollis* (Hymenoptera: Apidae). *Ethology.* in press.

Collins, A. M. 1979. Genetics of the response of the honey bee to an alarm chemical, isopentyl acetate. *J. Apicult. Res.* 18:285-291.

Contel, E. P. B., M. A. Mestriner and E. Martins. 1977. Genetic control and developmental expression of malate dehydrogenase in *Apis mellifera. Biochem. Genet.* 15:859-876.

Craig, R. 1980. Sex investment ratios in social hymenoptera. *Am. Nat.* 116:311-323.

Crozier, R. H. and P. C. Consul. 1976. Conditions for genetic polymorphism in social hymenoptera under selection at the colony level. *Theor. Pop. Biol.* 10:1-9.

Crozier, R. H. and R. E. Page. 1985. On being the right size: male contributions and multiple mating in social Hymenoptera. *Behav. Ecol. Sociobiol.* 18:105-115.

Crozier, R. H., P. Pamilo and Y. C. Crozier. 1984. Relatedness and microgeographical genetic variation in *Rhytidoponera mayri*, an Australian arid-zone ant. *Behav. Ecol. Sociobiol.* 5:143-150.

Del Lama, M. A., A. E. E. Soares and M. A. Mestriner. 1985. Linkage studies in *Apis mellifera* using biochemical and morphological markers. *J. Hered.* 76:427-430.

Fergusson, L. A. and M. L. Winston. The influence of wax deprivation on temporal polyethism in honey bee (*Apis mellifera* L.) colonies. *Can. J. Zool.* in press.

Frumhoff, P. C. and S. Schneider. 1987. The social consequences of honey bee polyandry: the effects of kinship on worker interactions within colonies. *Anim. Behav.* 35:255-262.

Frumhoff, P. C. and J. Baker. 1988. A genetic component to division of labour within honey bee colonies. *Nature* 333:358-361.

Gamboa, G. J., Reeve, H. K. and D. W. Pfennig. 1986. The evolution and ontogeny of nestmate recognition in social wasps. *Ann. Rev. Entomol.* 31:431-454.

Gary, N. E. 1960. A trap to quantitatively recover dead and abnormal honeybees from the hive. *J. Econ. Ent.* 53:782-785.

Gary, N. E. and K. Lorenzen. 1976. A method for collecting the honey-sac contents from honeybees. *J. Apicult. Res.* 15:73-79.

Jeanne, R. L. 1980. Evolution of social behavior in the Vespidae. *Ann. Rev. Entomol.* 25:371-396.

Jeanne, R.L., ed. 1988. *Interindividual behavioral variability in social insects.* Westview Press:Boulder, CO.

Jeanne, R. L., Downing, H. A. and D. C. Post. 1988. Age polyethism and individual variation in *Polybia occidentalis,* an advanced eusocial wasp. In *Interindividual behavioral variability in social insects,* R.L. Jeanne, ed. Westview Press:Boulder, CO. pp. 323-358.

Jones, R. J. 1980. Gallery construction by *Nasutitermes costalis*: polyethism and the behavior of individuals. *Insectes Sociaux* 27:5-28.

Kaftanoglu, O. and Y. S. Peng. 1980. A washing technique for collection of honeybee semen. *J. Apicult. Res.* 19:205-211.

Kolmes, S. A. 1985. An ergonomic study of *Apis mellifera* (Hymenoptera: Apidae). *J. Kans. Entomol. Soc.* 58:413-421.

Laidlaw, H. H. 1974. Relationships of bees within a colony. *Apiacta* 9:49-52.

Laidlaw, H. H. 1977. *Instrumental insemination of honey bee queens.* Dadant and Sons:Hamilton, Ill.

Laidlaw, H. H. and R. E. Page. 1984. Polyandry in honey bees (*Apis mellifera* L.): sperm utilization and intracolony genetic relationships. *Genetics* 108:985-997.

Lindauer, M. 1955. The water economy and temperature regulation of the honeybee colony. *Bee World* 35:62-72, 81-92, 105-111.

Metcalf, R. A. and G. S. Whitt. 1977a. Intra-nest relatedness in the social wasp *Polistes metricus.* A genetic analysis. *Behav. Ecol. Sociobiol.* 2:339-351.

Metcalf, R. A. and G. S. Whitt. 1977b. Relative inclusive fitness in the social wasp *Polistes metricus.. Behav. Ecol. Sociobiol.* 2:353-360.

Moore, A. J., M. D. Breed and M. J. Moor. 1987. The guard honey bee: ontogeny and behavioural variability of workers performing a specialized task. *Anim. Behav.* 35:1159-1167.

Moritz, R. F. A. 1983. Homogeneous mixing of honeybee semen by centrifugation. *J. Apicult. Res.* 24:249-255.

Moritz, R. F. A. 1985. The effects of multiple mating on the worker-queen conflict in *Apis mellifera* L. *Behav. Ecol. Sociobiol.* 16:375-377.

Moritz, R. F. A. and E. E. Southwick. 1987. Phenotype interactions in group behavior of honey bee workers (*Apis mellifera* L.). *Behav. Ecol. Sociobiol.* 21:53-57.

Nowogrodzki, R. 1983. *Individual differences and division of labor in honey bees.* Ph.D. Dissertation, Cornell University:Ithaca, New York.

Nowogrodzki, R. 1984. Division of labour in the honeybee colony: a review. *Bee World* 65:109-116.

Nunamaker, R. A. and W. T. Wilson. 1980. Some isozymes of the honey bee (*Apis mellifera* L.). *Isozyme Bull.* 13:111-112.

Oster, G. F. and E. O. Wilson. 1978. *Caste and ecology in the social insects.* Princeton University Press:Princeton, NJ.

Owen, R. E. 1986. Colony-level selection in the social insects: single locus additive and nonadditive models. *Theor. Pop. Biol.* 29:198-234.

Page, R. E. 1986. Sperm utilization in social insects. *Ann. Rev. Entomol.* 31:297-320.

Page, R. E. and M. D. Breed. 1987. Kin recognition in social bees. *Trends in Ecol. and Evol.* 2:272-275.

Page, R. E. and R. A. Metcalf. 1982. Multiple mating, sperm utilization, and social evolution. *Am. Nat.* 119:263-281.

Post, D. C., Jeanne, R. L. and E. H. Erickson. 1988. Variation in behavior among workers of the primitively eusocial wasp *Polistes fuscatus variatus.* In *Interindividual behavioral variability in social insects,* R.L. Jeanne, ed. Westview Press:Boulder, CO. pp. 283-322.

Ribbands, C. R. 1952. Division of labour in the honeybee community. *Proc. Roy. Soc. Lond. B.* 140:32-43.

Rinderer, T. E. and A. M. Collins. 1986. Behavioral genetics. In *Bee genetics and breeding,* T.E. Rinderer, ed. Academic Press:NY. pp. 155-176.

Robinson, G. E. 1987. Modulation of alarm pheromone perception in the honey bee: evidence for division of labor based on hormonally regulated response thresholds. *J. Comp. Physiol. A* 160:613-619.

Robinson, G. E. and R.E. Page. 1988. Genetic determination of guarding and undertaking in honey bee colonies. *Nature* 333:356-358.

Robinson, G. E., Strambi, A., and C. Strambi. 1988. Regulation of honey bee behavioral plasticity by juvenile hormone. In *Endocrinological frontiers in physiological insect ecology,* F. Sehnal, Z. Zabza, D.L. Denlinger, eds. Wroclaw Technical University Press:Wroclaw, Poland. pp. 691-694.

Rothenbuhler, W. C. 1964. Behavior genetics of nest cleaning in honey bees. IV. Responses of F_1 and backcross generations to disease-killed brood. *Am. Zool.* 4:111-123.

Sakagami, S. F. 1953. Untersuchung über die Arbeitsteilung in einem Zwergvolk der Honigbiene. Beiträge zur Biologie des Bienenvolkes, *Apis mellifera* L. I. *Jap. J. Zool.* 11:117-185.

Seeley, T. D. 1982. Adaptive significance of the age polyethism schedule in honeybee colonies. *Behav. Ecol. Sociobiol.* 11:287-293.

Seeley, T. D. 1985. *Honeybee ecology: A study of adaptation in social life.* Princeton University Press:Princeton, NJ.

Seeley, T. D and R. A. Levien. 1987. Social foraging by honeybees: how a colony tracks rich sources of nectar. In *Neurobiology and behavior of honeybees,* R. Menzel and A. Mercer, eds. Springer-Verlag, Berlin. pp. 266-279.

Sekiguchi, K. and S. F. Sakagami. 1966. Structure of foraging population and related problems in the honeybee, with considerations on the division of labour in bee colonies. *Hokk. Natl. Agric. Exp. Sta. Rep.* 69:1-65.

Sommeijer, M. J. 1983. *Social mechanisms in stingless bees.* Meppel:Utrecht, Holland.

Trivers, R. L. and H. Hare 1976. Haplodiploidy and the evolution of social insects. *Science* 191:249-263.

Visscher, P. K. 1983. The honey bee way of death: necrophoric behaviour in *Apis mellifera. Anim. Behav.* 31:1070-1076.

Wilson, E. O. 1971. *The insect societies.* Belknap Press of Harvard University Press:Cambridge, Mass.

Wilson, E. O. 1983. Caste and division of labor in leaf-cutter ants (Hymenoptera: Formicidae: *Atta*). *Behav. Ecol. Sociobiol.* 14:47-54.

Wilson, E. O. 1985. The sociogenesis of insect colonies. *Science* 228:1489-1495.

Winston, M. L. 1987. *The biology of the honey bee.* Harvard University Press:Cambridge, Mass.

Winston, M. L. and S. J. Katz. 1982. Foraging differences between cross-fostered honeybee workers (*Apis mellifera*) of European and Africanized races. *Behav. Ecol. Sociobiol.* 10:125-129.

CHAPTER 5

ECOLOGICAL DETERMINANTS OF SOCIAL EVOLUTION

Joan E. Strassmann and David C. Queller

The key evolutionary problem concerning the origin of eusociality, and of helping systems in general, is how genes could be selected when their effect is to decrease the reproduction of their bearers. Several theories attempt to account for cooperative behavior by denying that it is truly altruistic. Even though there may be some cost to the helping behavior, these theories posit some personal fitness gain that more than compensates for the cost. This compensatory gain may be an immediate, mutualistic one. For example, while joint defense against predators or parasites may carry some risks, it could cause a net increase in the personal fitness of all participants (Lin and Michener 1972). Alternatively, the gain may be delayed. Temporary helping may be favored if the helper eventually acquires a valuable territory (Emlen 1984, Woolfenden and Fitzpatrick 1984, Brown 1987) or reciprocal help comes at a later time (Trivers 1974, Emlen 1981). These theories may account for some cases of helping, but they do not explain truly altruistic behavior for which there is no compensating personal gain. In most eusocial insects there are no more than a few reproductives at a time, so the majority of individuals cannot gain any immediate personal advantage. Delayed benefits sometimes apply in eusocial insects: queen death may provide opportunities for others to reproduce. But this cannot be a general explanation for two reasons. First, the number of vacancies is low and the number of workers is large, so the average worker stands to gain very little. Second, in some species, the replacement egg-layers are individuals that did not help much, having focused instead on avoiding risks or establishing social dominance (e.g. Gadagkar and Joshi 1984).

Department of Biology, Rice University, P. O. Box 1892, Houston TX 77251, USA

Kin selection, indirect reproduction through helping relatives who tend to share the same genes, must provide the selective advantage for the many cases in which workers have little expectation of personal reproduction. Elements of the theory of kin selection were anticipated in various discussions of altruistic behavior (Darwin 1859, Fisher 1930, Haldane 1955, Williams and Williams 1957) as well as in the animal breeding literature (Lush 1947, Willham 1963). But it was Hamilton (1963 1964a 1964b 1970 1972) who combined a simple quantitative formulation of the idea with a recognition of its widespread importance in the natural world. Using Hamilton's assumptions one can write the condition necessary for selection to favor altruistic behavior as:

$$r^* \Delta w^* > r \Delta w . \qquad\qquad (1)$$

Here Δw is the number of offspring lost by the altruist and r is its relatedness to them (usually 1/2), while Δw^* is the number of offspring gained by the beneficiary, and r^* is the relatedness of the altruist to them.

Two classes of explanations of the evolution of eusociality emerge from Hamilton's rule. One focuses on differences in the r terms and the other on differences in the Δw's or costs and benefits. Both may operate together, but for purposes of clarity we discuss them separately. The next section discusses the role of relatedness in the evolution of eusociality and focuses on the haplodiploid hypothesis. The following sections subdivide hypotheses based on the costs and benefits side of Hamilton's rule into those invoking limited reproductive opportunities outside groups and those postulating ecological advantages to groups which result in higher inclusive fitness for group members. In each case, wasps of the genus *Polistes* are treated in detail because *Polistes* is one of the few genera (perhaps the only one) in which many of the hypotheses for group living and altruism have been investigated.

Polistes is especially appropriate for investigating the origin of eusociality because it is primitively social and lacks morphological castes. Its open-faced colonies are easy to find and study (FIGURE 1). This makes many of the key ecological parameters such as colony predation and parasitism much easier to document than is the case for the often cryptic colonies of ground and twig nesting bees and wasps (Michener 1964, Wilson

FIGURE 1. Colony of *Polistes exclamans.*

1971). [For a summary of the typical colony cycle of *Polistes* see Strassmann and Hughes (1986)].

RELATEDNESS AND THE EVOLUTION OF EUSOCIALITY

Besides devising the general inclusive fitness formula, Hamilton (1964a,b) developed a specific relatedness-centered hypothesis for the origin of eusociality. He noted that the haplodiploid genetic system possessed by this group leads to some peculiar relatedness properties. Specifically, a female who helps to rear super sisters would obtain a particularly favorable r^* value of 3/4, instead of the usual 1/2 for diploids (formula 1). Selection could then sometimes favor rearing super sisters instead of offspring even if $\Delta w^* < \Delta w$. This hypothesis, called the haplodiploid hypothesis, would account for the fact that most of the origins of eusociality occurred in the haplodiploid Hymenoptera and for the fact that workers in this group are exclusively female.

Though elegant, the haplodiploid hypothesis has encountered a series of problems. First, females are related to their brothers by only 1/4 (Crozier 1970), so that a haplodiploid female helping to rear an even mixture of super sisters and brothers would obtain an average r to brood of only 1/2, exactly the same value as in diploid systems. The haplodiploid advantage could be saved if female workers could avoid aiding brothers, either because of a female-biased sex ratio or by laying the male-destined eggs themselves (Hamilton 1972, Trivers and Hare 1976). Worker laying of male-destined eggs occurs, but is far from universal (Hamilton 1972, Trivers and Hare 1976, Bourke, 1988). Female-biased sex ratios also occur (Trivers and Hare 1976, Nonacs 1986), but their significance in selecting for eusociality is not entirely clear for two reasons. First, once workers succeed in biasing the sex ratio to their optimum (3 females : 1 male under the simplest conditions), the haplodiploid advantage disappears because the rarer males become more valuable than females (Trivers and Hare 1976, Grafen 1986). The maintenance of eusociality must therefore be explained in some other way. Second, there is also a problem for the origin of eusociality because it requires two simultaneous changes: helping and the female-biased sex ratio (Charnov 1978). Still the haplodiploid hypothesis can be rescued under certain conditions. Seger (1983) and Grafen (1986) have described circumstances in which a female-biased sex ratio will be favored among the progeny of only some mothers, setting up the opportunity for their daughters to stay and rear mainly sisters, without females being devalued relative to males.

While these adjustments to the haplodiploid hypothesis make it more logically sound, it is still questionable whether it actually describes the critical condition for the origin and maintenance of eusociality (Lin and Michener 1972, Alexander 1974, Evans 1977, West-Eberhard 1975 1978, Eickwort 1981, Brockmann 1984, Andersson 1984, Stubblefield and Charnov 1986, Alexander and Noonan *in press*). First, it cannot account for the diploid termites. Some special mechanisms could increase relatedness in this group (Luykx and Syren 1979, Bartz 1979, Lacy 1980 1984, Pamilo 1984), but it is not clear that these mechanisms operated in the early termites (Leinaas 1983, Crozier and Luykx 1985). Nor can the relatedness hypothesis account for helping in diploid vertebrates, from the apparently eusocial naked mole rats (Jarvis 1981) to the many species with less developed social systems (Vehrencamp 1979, Emlen 1984, Brown 1987).

It is also far from clear that the haplodiploid hypothesis explains eusociality even in the Hymenoptera. The value of r = 3/4 depends on having one singly-mated egg layer per colony. Multiple mating is widespread in insects (Parker 1970, Thornhill and Alcock 1983) including some eusocial Hymenoptera (reviewed by Page 1986), though one can only speculate about the mating habits of the ancestral species in which eusociality first appeared. Multiple egg laying individuals are also widespread in eusocial colonies (Wilson 1971, Jeanne 1980). Moreover, it is probable that queen death led to multiple (sequential) egg layers in the earliest eusocial species (West-Eberhard 1978); prior to having evolved special longevity, a mother would usually die before her daughters, leaving the nest to another egg layer. Thus sociality probably first evolved with an average relatedness among female brood of less than 3/4. Whatever the theoretical arguments, measurements of relatedness make it clear that eusociality is maintained in some species with relatedness to female brood of less than 1/2 (Craig and Crozier 1979, Pamilo and Varvio-Aho 1979, Ward and Taylor 1981, Pamilo 1982a 1982b, Pearson 1982, Ward 1983, Crozier *et al.* 1984, Ross and Fletcher 1985, Ross 1986).

Two separate studies of *Polistes exclamans* found relatedness values between workers and brood of about 0.38, which is lower than worker relatedness to their own young (Lester and Selander 1981, Strassmann 1985b). These are the only such studies in *Polistes*. It is especially interesting that workers usually remain on their natal nest in this species in spite of low relatedness as workers of *P. exclamans* sometimes leave their natal nest to begin a new (satellite) nest of their own (Strassmann 1981a). The decisions of other workers to join such a satellite nest depend on their relatedness to the queen of the satellite; they join her if their relatedness is high (Strassmann 1981b). These results indicate that workers sometimes respond to alternatives, make choices based on relatedness, and yet usually choose to remain on their natal nest in spite of the presence of an alternative (nesting alone) which would yield higher relatedness to the brood. Coming back to Hamilton's rule, the available evidence on relatedness indicates that ecological factors must be important in the origin and maintenance of eusociality.

COSTS AND BENEFITS OF EUSOCIALITY

Perhaps because of the limitations of the haplodiploid hypothesis, current opinion seems to favor explanations that stress costs and benefits (Lin and Michener 1972, Alexander 1974, Evans 1977, West-Eberhard 1975 1978, Eickwort 1981, Brockmann 1984, Andersson 1984, Stubblefield and Charnov 1986, Alexander and Noonan *in press*). If there is no special relatedness advantage, then the best case (in an outbreeder) is helping full sibs: $r^* = r = 1/2$. This means that even in the best of circumstances eusociality cannot be favored unless the benefit of helping is greater than the cost, $\Delta w^* > \Delta w$ (see formula 1).

Therefore, at the very least, we need to explain how an individual can rear more of a relative's offspring that it can rear of its own. Compared to the theoretical work on relatedness, this area has been relatively neglected. A few theoretical studies have addressed the question of how the organization of work within colonies can evolve to make the colony more efficient (Oster and Wilson 1978, Wilson 1985, Jeanne 1986), but in the early stages of eusociality, efficient work patterns would not yet have evolved. Moreover, it is not clear that these mechanisms always give the required efficiency. From collections of colonies of social insects, Michener (1964) found that *per capita* productivity declines with colony size in most species. This suggests that the most productive colony size is one, and that each worker adds less to the colony than it could have reared on its own; the $\Delta w^* > \Delta w$ requirement is not met. There are, however, two general ways of reconciling the observation with the requirement that $\Delta w^* > \Delta w$. The first hypothesizes differences in reproductive potential among individuals, and is commonly referred to as 'best of a bad job' (e.g. Brockmann 1984). The second postulates fitness gains to individuals in groups that would not be detected in studies like Michener's (1964). This is where true ecological advantages to grouping fall. These two categories of costs-and-benefits explanations for eusociality will be discussed in the next two sections.

BEST OF A BAD JOB AND THE EVOLUTION OF EUSOCIALITY

Extrinsic or intrinsic constraints may limit an individual's reproductive ability when it nests alone. Extrinsic constraints include shortages of nesting sites or territories, shortages of mates, and shortages of time for nesting (e.g. Brown 1987, Herbers 1986, Strassmann *et al.* 1987). Intrinsic constraints make some females less capable of laying eggs than they are of tending another female's brood (Craig 1983). Both types of constraints can result in selection on females to join groups rather than nest alone. They are both consistent with Michener's finding of decreasing *per capita* fitness with group size because some individuals may be incapable of nesting alone, so that even if they add relatively little to the colony success, they are still opting for the best alternative in a bad situation. These are discussed below with special reference to *Polistes*.

Nest Site Constraints

One of the most important requirements for any organism that displays parental care is to have a safe site in which to raise the young. This is particularly important when the young are immobile and defenseless, as is the case in altricial birds (e.g. Brown 1987), many carnivores (e.g. Moehlman 1986), and most insects that exhibit parental care (e.g. Eickwort 1981). If a particular feature of the environment renders it more protected than other areas, then the number of such locations may limit the numbers of individuals that can raise young. If all safe sites have been filled, an individual may achieve higher inclusive fitness by helping a relative rear additional progeny in a safe location than it could raise by starting its own nest in an unsafe location.

Demonstrations that this principle is important in group nesting require identification of the limiting habitat, demonstration that all sites are filled, and observation that non-breeders can help breeders in the safe sites. Experiments can verify that group size diminishes when additional nest sites are added. To our knowledge the only study of this was conducted by Herbers (1986) on *Leptothorax longispinosus*. Polygyny (queens sharing a nest) decreased when nest sites were added to the acorns and hollow twigs usually used for nests. Other examples of limited nest sites are caves, holes in the ground and hollow trees. The re-use of natal nest sites even when facing the disadvantages of parasites that have accumulated in such sites also suggests that nest sites are limited. This is common in a variety of social insects (e.g. Batra 1966, Starr 1975, Evans and Hook 1982, Strassmann 1983, Kukuk and Eickwort 1987, Michener 1985, Itô *et al.* 1985, Hook 1985 1987). In fact, re-use of the natal nest or the natal nest site is a common way for groups to begin in primitively social bees and wasps. Re-use of the natal nest may not fall under best-of-a-bad-job conditions if it is the simplest way of finding relatives (which then allow individuals to obtain ecological advantages), or if there is a substantial saving in time and expense necessary to construct a new nest.

There is no good evidence that nest site constraints are important factors selecting for grouping in *Polistes*. Many species nest in vegetation, on cliffs, and on eaves where nest sites seem to be nearly infinite (e.g. West-Eberhard 1969, Strassmann 1979, Strassmann and Hughes 1986). It may be argued that space is limited on a cliff or eave (Noonan 1979, Cervo and Turillazzi 1985, Queller and Strassmann 1988). However, such arguments are susceptible to the criticism that the nests could have been packed more closely together. *P. carolinus* may exemplify grouping because of nest site constraints as it nests in dark places like artificial nest boxes and hollow trees. This species uses the same nest boxes year after year and prepares them for re-use by tearing down the nest at the end of the season (Hughes and Strassmann, unpublished). Nests are begun by 2.2 foundresses on average (the range is 1 to 7), a larger average number of foundresses than other species in the area (Hughes and Strassmann in press). However, average foundress numbers per nest did not change from 1982 to 1987 even though numbers of new nest boxes varied annually. Thus, the addition of nest boxes does not cause the decrease in foundress association size that would be expected if females group because of the lack of suitable nest sites.

Habitat constraints, which are in many respects similar to nest site constraints, are favorite explanations for helpers at the nest in many bird species (e.g. Koenig and Mumme 1987, Woolfenden and Fitzpatrick 1984). While habitat is certainly restricted in a number of bird species with helpers, it is not clear that this is the driving force behind helping (Brown 1987, Austad and Rabenold 1985). For example, Koenig and Mumme (1987) noted that adding or removing granaries did not have a marked effect on group size in acorn woodpeckers.

Subfertility

A related best-of-a-bad-job hypothesis supposes that some individuals are in poor condition (or "subfertile"; West-Eberhard 1978) such that they are unlikely to succeed at reproducing alone. These individuals could make the best of their situation by helping, provided that they are not also handicapped in their abilities to help (Craig 1983). In some cases, the poor condition may be imposed by parental manipulation (Alexander 1974) in order to force some progeny to be helpers. This kind of hypothesis must certainly apply to many higher eusocial insects with differentiated castes. Workers are less capable of being productive egg layers, indeed sometimes completely incapable. However, attempts to find support for this hypothesis in social insects without distinct castes have failed (Haggard and Gamboa 1980, Sullivan and Strassmann 1984, Kukuk and Eickwort 1987, Queller and Strassmann, this volume). This failure is serious if the hypothesis is to explain the origin of eusociality as eusociality presumably evolved prior to the development of castes. But the failure should perhaps come as no surprise since the hypothesis requires the uncoupling of talents for foraging from those for egg laying (Craig 1983).

The subfertility hypothesis has been accepted as pivotal in the evolution of eusociality in part because it is so logical and attractive an explanation. Here we summarize the results of our attempts to find support for the hypothesis in *P. annularis* (Strassmann 1979, Sullivan and Strassmann 1984, Queller and Strassmann 1988). We also addressed the subfertility hypothesis in *P. bellicosus* (Queller and Strassmann, this volume).

According to the subfertility hypothesis, poor quality females that are not capable of nesting alone should join relatives on nests where they will function as subordinates. In *P. annularis* we measured female quality in two ways: size (Sullivan and Strassmann 1984, Strassmann 1983) and whether or not they had been deprived of honey during the winter (Strassmann 1979). Both measures had significant effects on female fitness. However, neither small females nor females that had been deprived of honey were found in larger groups. Thus grouping was not a response of females to their overall lowered condition. The only study that seems to support the subfertility hypothesis demonstrated that under laboratory conditions some females of *P. fuscatus* failed to initiate nests (Gibo 1974). However even females of singly-founded species frequently fail to build nests in the laboratory because of the artificial conditions, so this experiment cannot be regarded as support for the subfertility hypothesis.

When females are infected with stylopid parasites, reproduction is reduced more drastically than is the ability to work, as the first effect of the parasite is to sterilize the female (Strambi 1965, Strambi and Strambi 1973). Selection might favor such females working on a sister's nest because the female could work but not lay eggs. But she would probably be better off staying away from all relatives as she might pass on the parasites to the brood she reared, thus having a net negative effect on the relative's progeny. Also stylopid parasites are not common enough to account for most workers.

Therefore, we conclude that there is no evidence that either nest site constraints or subfertility have been important in the evolution of eusociality. However there have been so few empirical tests of either hypothesis that it is entirely possible that this position could change with the publication of new studies that focus on these questions, particularly on previously unstudied taxa.

Ecological Advantages of Eusociality

If the best of a bad job explanation fails, then group members should have higher inclusive fitness than solitary individuals. Solitary individuals may occur in species that ordinarily nest in groups if they cannot find any relatives with whom to nest (Noonan 1981). A comprehensive study of all individuals in a given area from colony inception to the production of reproductives should reveal the advantages of group nesting. Detailed studies of the correlates of reproductive success should also reveal the specific selective factors that give groups their special advantages over solitary individuals. This is particularly likely if the study is conducted in native habitats where the selective factors for grouping are likely to be operating.

Unfortunately, relatively few such studies have been carried out on primitively eusocial insects (e.g. Batra 1966, Jeanne 1972, Sakagami 1977, Litte 1977, 1979, Noonan 1979, 1981, Pickering 1980, Strassmann 1981c, Queller and Strassmann 1988, Strassmann in preparation, Strassmann and Hughes in preparation). These studies indicate that the probability that a colony will fail is critical in assessing the costs and benefits of grouping. Smaller groups of females are, in general, more likely to fail in their nesting attempts than are larger groups of females. Failure may be due to loss of all tending adults, or to loss of the nest to predation and the subsequent failure to build a new nest successfully. This advantage of larger groups would not be detected in a study in which nests were collected at one time because the failed nests would not be represented in the collection. Following colonies over longer periods of time is necessary to assess the importance of predation on adults and brood, as was noted by Michener (1964). Predation on brood in nests and predation on adults may confer an advantage to larger groups.

Colony Defense

Brood may be killed individually by parasitoids or all at once by a nest predator which removes the entire nest and consumes the contents. Larger associations may be better protected from predators and parasitoids due to increased vigilance, or more effective stinging attacks (Alexander 1974). Defense against parasitoids is often cited as an advantage of early grouping in colonies (e. g. Evans 1977). Such an advantage should translate into greater *per capita* fitness for females in groups and could be detected simply by collecting colonies. Since such collections do not usually indicate that there is a *per capita* advantage (Michener 1964), this is unlikely to be the key factor favoring grouping. In contrast, increased abilities for defense of the entire nest from destruction by a predator would not be detected by a collection, and may be a major advantage to group nesting (Alexander 1974). However, we are unaware of any study that has documented lower predation rates for colonies defended by greater numbers of primitively eusocial insects. In *Polistes* there is no evidence that larger groups deter parasitoids more effectively (Strassmann 1981c). Nor is there any evidence in this genus that larger groups of individuals are more effective at predator deterrence (Gibo 1978, Strassmann 1981c, Strassmann *et al.* 1988).

Even if the group is not successful at defense, it may be better at building a new nest after the original nest is destroyed. This is the case in *Polistes bellicosus* in which larger groups of individuals were better able to build a new nest and successfully rear brood in it (Strassmann *et al.* 1988). We removed 49 nests and their brood without harming the adults and then monitored the subsequent behavior and reproductive success of the adults (Strassmann *et al.* 1988). After 44 days those females which were originally in groups of 4 or more had much higher *per capita* fitness rates than did females in smaller groups, because the rebuilt nest was less likely to fail. In fact, among successful nests there was no relationship between numbers of adults produced after 44 days and group size. Including the failed nests in the analysis resulted in a *per capita* advantage to larger groups. As nests are destroyed very frequently in this population of *P. bellicosus*, it is critical to include nest failure in any analysis of the advantage of group size. The key advantage that we observed would not be detected in a simple collection of nests and their associated brood. In summary, groups are better able to recover from predation events than are solitary individuals. No such advantage has been detected for defense against either predators or parasitoids. Therefore it is critical that studies of the costs and benefits of group nesting follow colonies long enough to measure the effects of nest predation.

Demographic Advantages

One disadvantage of trying to rear young alone is that death of the parent leads to the death of all its dependent young. This may be particularly severe for insects in which high adult mortality rates are combined with extended periods of complete dependency of young. For example, in the wasps *Polistes exclamans* (Strassmann and Orgren 1983, Strassmann 1985a) and *Mischocyttarus drewseni* (Jeanne 1972) the average adult worker longevity as an adult is considerably less than the time required to rear an offspring to independence. Most individuals, if they nested alone, would therefore fail completely. Cooperative care of the young can help circumvent this problem in two related ways.

First, when a worker or a co-foundress dies, its investment in still-dependent young is protected by the presence of other adults who can rear the young to the age of independence. Thus the investment in brood of a group-nesting individual is insured against that individual's death by the presence of others that would continue to rear the young. In *Polistes* having at least one foundress survive to the worker emergence stage is an important advantage of foundress associations (Metcalf and Whitt 1977, Queller and Strassmann 1988). Even though colonies tend to be larger later in the colony cycle, the same advantage probably also applies to workers remaining on the natal nest. Even in mid-summer the main cause of colony failure in *Polistes exclamans* is the death or disappearance of all adults (Strassmann 1981c). Larger colonies are less likely to fail for this reason than smaller ones.

There is a related, but distinct, advantage of cooperative brood care in the face of high adult mortality and extended parental care. Besides having another adult to carry on one's investment after one's death, it is also possible to carry on investment that others have started (Queller submitted). This is not necessarily the same as the first advantage. For example, a worker staying on its natal nest can obtain an advantage by carrying on the queen's initial brood investment even if the queen is still alive. The advantage to the worker is a reproductive head start. If it had nested alone, it would have to wait at least the length of the entire developmental period and run the risk of dying and failing to realize any of its investment. As a worker, however, it can begin to realize some of its reproductive investment very quickly because it can invest in young that are already partly grown. If it is unlucky enough to suffer an early death, it will still have realized some of its reproductive investment. Analysis of demographic data on four species of polistine wasps shows this to be a strong advantage favoring worker behavior (Queller submitted). Termite workers have a special head start advantage as juveniles can function as workers but cannot yet reproduce on their own (Alexander and Noonan in press).

In summary, the combination of high mortality with an extended period of parental care makes it unlikely that any given adult will successfully rear offspring on its own. Cooperation is advantageous under these conditions because an individual can successfully reproduce even if it survives for a relatively short period. This kind of advantage would not be evident in Michener's (1964) studies of colony productivity because one-time collections give no information about individuals that have died. Further research on the demography of social insects is necessary to determine the generality and magnitude of this factor in selecting for group living.

CONCLUSIONS

If Evans' (1977) plea for more empirical studies of the insects important in the evolution of eusociality had been heeded, now would be the time for a definitive review. Instead we review a small list of field studies with predominantly negative results. Cooperating individuals are usually related to each other, but in most species worker to brood relatedness is not higher than worker to progeny relatedness. To date, neither subfertility or nest site constraints have been demonstrated to have played a major role in the early evolution of eusociality. By a process of elimination we conclude that there must be real ecological advantages to group nesting. This conclusion seems to be supported by the enormous ecological success of social insects (reviewed in Wilson 1987, Jeanne and Davidson 1984). Collections of colonies reveal a declining *per capita* output with group size but group living persists. Therefore, all indicators point to colony failure as a critical variable in the evolution of eusociality. Groups are protected against failure by their enhanced abilities to recover after loss of both nest and brood. Groups can also protect against the loss of all adults and subsequent starvation of the brood. Future studies would do well to focus on colony failure in looking for advantages to eusociality. However the number of empirical studies of native populations is so small that additional studies of relatedness levels, subfertility, or nest site constraints in appropriate populations may change this focus.

Acknowledgements

We thank Colin Hughes for collaborating on many of the cited field studies which were conducted at Brazos Bend State Park under Texas Parks and Wildlife permit numbers 15-82, 11-83, 8-84, 16-85, 20-86 and 18-87. We thank D. Riskind of the Texas Parks and Wildlife Department and L. Fowler and C. VanBaarle of Brazos Bend State Park for their cooperation. This research was partially supported by NSF BSR86-05026.

LITERATURE CITED

Alexander, R. D. 1974. The evolution of social behavior. *Ann. Rev. Ecol. Syst.* 4:325-383.

Alexander, R. D. and K. M. Noonan. In press. The evolution of sociality. In *The Natural History and Social Behavior of Naked Mole Rats*, P. W. Sherman, J. Jarvis and R. D. Alexander, eds.

Andersson, M. 1984. The evolution of eusociality. *Ann. Rev. Ecol. Syst.* 15:165-189.

Austad, S. N. and K. N. Rabenold 1985. Reproductive enhancement by helpers and an experimental inquiry into its mechanism in the bicolored wren. *Behav. Ecol. Sociobiol.* 17:19-27.

Bartz, S. 1979. Evolution of eusociality in termites. *Proc. Natl. Acad. Sci. USA* 76:5764-5768.

Batra, S. W. T. 1966. The life cycle and behavior of the primitively social bee, *Lasioglossum zephyrum* (Halictidae). *Univ. Kansas Science Bulletin* 46:359-423.

Bourke, A. F. G. 1988. Worker reproduction in the higher eusocial Hymenoptera. Q. Rev. Biol. 63:291-311.

Brockmann, H. J. 1984. The evolution of social behavior in insects. In *Behavioural Ecology, An Evolutionary Approach*, J. R. Krebs and N. B. Davies, eds. Second Edition. Sinauer Assoc. Inc.:Sunderland, Mass. pp. 340-361.

Brown, J. L. 1987. *Helping in Communal Breeding Birds: Ecology and Evolution.* Princeton University Press:Princeton, N.J.

Cervo, R. and S. Turillazzi. 1985. Associative foundation and nesting sites in *Polistes nimpha. Naturwissenschaften* 72:48-49.

Charnov, E. L. 1978. Evolution of eusocial behavior: offspring choice or parental parasitism. *J. Theor. Biol.* 75:451-465.

Craig, R. 1983. Subfertility and the evolution of eusociality by kin selection. *J. Theor. Biol.* 100:379-397.

Craig, R. and R. H. Crozier. 1979. Relatedness in the polygynous ant *Myrmecia pilosula. Evolution* 33:335-341.

Crozier, R. H. 1970. Coefficients of relationship and the identity of genes by descent in the Hymenoptera. *Am. Nat.* 104:216-217.

Crozier, R. H., P. Pamilo and Y. C. Crozier. 1984. Relatedness and microgeographic genetic variation in *Rhytidoponera mayri*, an Australian arid zone ant. *Behav. Ecol. Sociobiol.* 15:143-150.

Crozier, R. H. and P. Luykx. 1985. The evolution of termite eusociality is unlikely to have been based on male-haploid analogy. *Am. Nat.* 126:867-869.

Darwin, C. 1859. *The Origin of Species by Means of Natural Selection or The Preservation of Favored Races in The Struggle for Life.* Modern Library:New York.

Eickwort, G. C. 1981. Presocial insects. In *Social Insects*, H.R. Herman,ed. Academic Press, New York pp.199-280.

Emlen, S. T. 1981. Altruism, kinship, and reciprocity in the white-fronted bee-eater. In *Natural Selection and Social Behavior*, R. D. Alexander and D. W. Tinkle, eds. Chiron Press:New York pp. 217-230.

Emlen, S. T. 1984. Cooperative breeding in birds and mammals. In *Behavioural Ecology, An Evolutionary Approach*, J. R. Krebs and N. B. Davies, eds. Sinauer Assoc. Inc.:Sunderland, Mass. pp. 305-339.

Evans, H. E. 1977. Extrinsic versus intrinsic factors in the evolution of insect sociality. *BioScience* 27:613-617.

Evans, H. E. and A. W. Hook. 1982. Communal nesting in the digger wasp, *Cerceris australis* (Hymenoptera: Sphecidae) *Aust. J. Zool.* 30:557-568.

Fisher, R. A. 1930. *The Genetical Theory of Natural Selection.* Clarendon Press:Oxford. (Reprinted and revised, 1958. Dover:New York).

Gadagkar, R. and N. V. Joshi. 1984. Social organization in the Indian wasp *Ropalidia cyathiformis* (Fab.) (Hymenoptera: Vespidae). *Z. Tierpsychol.* 64:15-32.

Gibo, D. L.   1974.   A laboratory study on the selective advantage of foundress associations in *Polistes fuscatus* (Hymenoptera: Vespidae).   *Can. Entomol.*   106:101-106.

Gibo, D. L.   1978.   The selective advantage of foundress associations in *Polistes fuscatus* (Hymenoptera: Vespidae): a field study of the effects of predation on productivity.   *Can. Entomol.*   110:519-540.

Grafen, A.   1986.   Split sex ratios and the evolutionary origins of eusociality.   *J. Theor. Biol.*   122:95-121.

Haggard, C. M. and G. J. Gamboa.   1980.   Seasonal variation in body size and reproductive condition of a paper wasp, *Polistes metricus* (Hymenoptera: Vespidae).   *Can. Ent.*   112:101-106.

Haldane, J. B. S.   1955.   Population genetics.   *New Biol.*   18:34-51.

Hamilton, W. D.   1963.   The evolution of altruistic behavior.   *Am. Nat.*   97:354-356.

Hamilton, W. D.   1964a.   The genetical evolution of social behaviour, I.   *J. Theor. Biol.*   7:1-16.

Hamilton, W. D.   1964b.   The genetical evolution of social behaviour, II.   *J. Theor. Biol.*   7:17-52.

Hamilton, W. D.   1970.   Selfish and spiteful behavior in an evolutionary model.   *Nature*   228:1218-1220.

Hamilton, W. D.   1972.   Altruism and related phenomena, mainly in the social insects.   *Ann. Rev. Ecol. Syst.*   3:193-232.

Herbers, J. M.   1986. Nest site limitation and facultative polygyny in the ant *Leptothorax longispinosus*. Behav. Ecol. Sociobiol. 19:115-122.

Hook, A. W.   1985.   *The comparative ethology of communal nesting in the Philanthinae (Hymenoptera:Sphecidae).* Ph. D. Dissertation, Colorado State Univ.:Fort Collins, Colorado   142 pp.

Hook, A. W.   1987.   Nesting behavior of Texas *Cerceris* digger wasps with emphasis on nest reutilization and nest sharing (Hymenoptera:Sphecidae).   *Sociobiology* 13:93-118.

Hughes, C. R. and J. E. Strassmann.   Foundress mortality after worker emergence in social wasps (*Polistes*) *Ethology* in press.

Itô, Y.,O. Iwahashi, S. Yamane and Sk. Yamane   1985.   Overwintering and nest reutilization in *Ropalidia fasciata* (Hymenoptera: Vespidae).   *Kontyû*   53:486-490.

Jarvis, J. V. M.   1981.   Eusociality in a mammal: Cooperative breeding in naked mole-rat colonies.   *Science*   212:571-573.

Jeanne, R. L.   1972.   Social biology of the neotropical wasp *Mischocyttarus drewseni*. *Bull. Mus. Comp. Zool. Harvard Univ.*   144:63-150.

Jeanne, R. L. 1980. Evolution of social behavior in the Vespidae. *Ann. Rev. Entomol.* 25:371-396.

Jeanne, R. L. 1986. The evolution of the organization of work in social insects. *Monitore zool. ital. (N.S.)* 20:119-133.

Jeanne, R. L. and D. W. Davidson. 1984. Population regulation in social insects. In *Ecological Entomology*, C. B. Huffaker and R. L. Rabb, eds. John Wiley and Sons:New York pp. 559-587.

Koenig, W. D. and R. L. Mumme. 1987. *Cooperatively breeding acorn woodpecker*. Princeton Univ. Press:Princeton, NJ. 435pp.

Kukuk, P. F. and G. C. Eickwort. 1987. Alternative social structures in halictine bees. In *Chemistry and Biology of Social Insects*, J. Eder and H. Rembold, eds. Verlag J. Peperny:Munich. pp. 555-556.

Lacy, R. C. 1980. The evolution of eusociality in termites: a haplodiploid analogy? *Am. Nat.* 116:449-451.

Lacy, R.C. 1984. The evolution of termite eusociality: reply to Leinaas. *Am. Nat.* 123:876-878

Leinaas, H. P. 1983. A haplodiploidy analogy in the evolution of termite eusociality? A reply to Lacy. *Am. Nat.* 121:302-304.

Lester, L. J. and R. K. Selander. 1981. Genetic relatedness and the social organization of *Polistes* colonies. *Am. Nat.* 117:147-176.

Lin, N., and C. D. Michener. 1972. Evolution of sociality in insects. *Quart. Rev. Biol.* 47:131-159.

Litte, M. 1977. Behavioral ecology of the social wasp, *Mischocyttarus mexicanus*. *Behav. Ecol. Sociobiol.* 2:229-246.

Litte, M. 1979. *Mischocyttarus flavitarsis* in Arizona: social and nesting biology of a polistine wasp. *Z. Tierpsych.* 50:282-312.

Lush, J. L. 1947. Family merit and individual merit as bases for selection. *Am. Nat.* 81: 241-261, 362-379.

Luykx, P., and R. M. Syren. 1979. The cytogenetics of *Incistitermes schwarzi* and other Florida termites. *Sociobiology* 4:191-209.

Metcalf, R. A. and G. S. Whitt. 1977. Relative inclusive fitness in the social wasp, *Polistes metricus*. *Behav. Ecol. Sociobiol.* 2:353-360.

Michener, C. D. 1964. Reproductive efficiency in relation to colony size in hymenopterous societies. *Insectes Sociaux* 11:317-341.

Michener, C. D. 1985. From solitary to eusocial: need there be a series of intervening species? In: *Experimental Behavioral Ecology*, B. Hölldobler and M. Lindauer, eds. Stuttgart pp. 293-305.

Moehlman, P. D. 1986. Ecology of cooperation in canids. In *Ecological aspects of social evolution*, D. I. Rubenstein and Richard W. Wrangham, eds. Princeton Univ. Press.:Princeton, NJ pp 64-86.

Nonacs, P. 1986. Ant reproductive strategies and sex allocation theory. *Quart. Rev. Biol.* 61:1-21.

Noonan, K. M. 1979. *Individual strategies of inclusive fitness maximizing in the social wasp, Polistes fuscatus (Hymenoptera:Vespidae).* Ph.D. Dissertation University of Michigan:Ann Arbor, Michigan.

Noonan, K. M. 1981. Individual strategies of inclusive-fitness-maximizing in *Polistes fuscatus* foundresses. In *Natural selection and social behavior*, R. D. Alexander and D. W. Tinkle, eds. Chiron Press:New York pp. 18-24.

Oster, G. F., and E. O. Wilson. 1978. *Caste and Ecology in the Social Insects.* Princeton University Press:Princeton, NJ

Page, R. E., Jr. 1986. Sperm utilization in social insects. *Ann. Rev. Entomol.* 31:297-320.

Pamilo, P. 1982a. Multiple mating in *Formica* ants. *Hereditas* 97:37-45.

Pamilo, P. 1982b. Genetic population structure in polygynous *Formica* ants. *Heredity* 48:95-106.

Pamilo, P. 1984. Genetic relatedness and evolution of insect sociality. *Behav. Ecol. Sociobiol.* 15:241-248.

Pamilo, P., and S.-L. Varvio-Aho. 1979. Genetic structure of nests of the ant *Formica sanguinea*. *Behav. Ecol. Sociobiol.* 6:91-98.

Parker, G. A. 1970. Sperm competition and its evolutionary consequences in the insects. *Biol. Rev. Cambridge Philos. Soc.* 45:525-567.

Pearson, B. 1982. Relatedness of normal queens (macrogynes) in nests of the polygynous ant *Myrmica rubra* Latreille. *Evolution* 36:107-112.

Pickering, J. 1980. *Sex ratio, social behavior and ecology in Polistes (Hymenoptera: Vespidae), Pachysomoides (Hymenoptera: Ichneumonidae) and Plasmodium (Protozoa, Haemosporida).* Ph. D. Dissertation, Harvard Univ.:Cambridge, Mass.

Queller, D. C. The evolution of eusociality: a head-start hypothesis. Submitted.

Queller, D. C., and J. E. Strassmann. 1988. Reproductive success and group nesting in the paper wasp, *Polistes annularis*. In *Reproductive Success: Studies in Individual Variation in Contrasting BreedingSystems*, T. H. Clutton-Brock, ed. University of Chicago Press:Chicago. pp. 76-96.

Ross, K. G. 1986. Kin selection and the problem of sperm utilization in social insects. *Nature* 323:798-800.

Ross, K. G., and D. J. C. Fletcher. 1985. Comparative study of genetic and social structure in two forms of the fire ant *Solenopsis invicta* (Hymenoptera: Formicidae). *Behav. Ecol. Sociobiol.* 17:349-356.

Seger, J. 1983. Partial bivoltinism may cause alternating sex-ratio biases that favour eusociality. *Nature* 301:59-62.

Sakagami S. F. 1977. Seasonal change of nest survival and related aspects in an aggregation of *Lasioglossum duplex* (Dalla Torre), a eusocial halictine bee (Hymenoptera: Halictidae). *Res. Popul. Ecol.* 19:69-86.

Starr, C.K. 1975. Nest reutilization by *Polistes metricus* (Hymenoptera:Vespidae) and possible limitation of multiple foundress associations by parasitoids. *J. Kansas Entomol. Soc.* 49:142-144.

Strambi, A. 1965. Influence du parasite *Xenos vesparum* Rossi (Strepsiptère) sur la neurosécrétion des individus du sexe femelle de *Polistes gallicus* L. (Hyménoptère, Vespidae). *C.R. Acad. Sci. Paris* 260:3767-3769.

Strambi, A. and C. Strambi. 1973. Influence du développement du parasite *Xenos vesparum* Rossi (Insecte, Strepsitère) sur le système neuroendocrinien des femelles de *Polistes* (Hyménoptère, Vespidae) au début de leur vie imaginale. *Arch. d'anotomie microscopique et de morphologie expérimentale* 62:39-54.

Strassmann, J. E. 1979. Honey caches help female paper wasps (*Polistes annularis*) survive Texas winters. *Science* 204:207-209.

Strassmann, J. E. 1981a. Evolutionary implications of early male and satellite nest production in *Polistes exclamans* colony cycles. *Behav. Ecol. Sociobiol.* 8:55-64.

Strassmann, J. E. 1981b. Kin selection and satellite nests in *Polistes exclamans*. In *Natural selection and social behavior*, R. D. Alexander and D. W. Tinkle, eds. Chiron Press:New York pp. 45-58.

Strassmann, J. E. 1981c. Parasitoids, predators and group size in the paper wasp, *Polistes exclamans. Ecology* 62:1225-1233.

Strassmann, J. E. 1983. Nest fidelity and group size among foundresses of *Polistes annularis* (Hymenoptera: Vespidae). *J. Kans. Entomol. Soc.* 56:621-634.

Strassmann, J. E.   1985a.   Worker mortality and the evolution of castes in the social wasp, *Polistes exclamans*. *Insectes Sociaux* 32:275-285.

Strassmann, J. E.   1985b.   Relatedness of workers to brood in the social wasp, *Polistes exclamans* (Hymenoptera: Vespidae). *Z. Tierpsychol.* 69:141-148.

Strassmann, J. E. and C. R. Hughes.   1986.   Latitudinal variation in protandry and protogny in polistine wasps. *Monitore zool. ital. (N.S.)* 20:87-100.

Strassmann, J. E. and Orgren, M. C.   1983.   Nest architecture and brood development times in the paper wasp, *Polistes exclamans* (Hymenoptera: Vespidae). *Psyche* 90:237-248

Strassmann, J. E., D. C. Queller and C. R. Hughes.   1987. Constraints on independent nesting by *Polistes foundresses* in Texas. In *Chemistry and biology of social insects*, J. Eder and H. Rembold, eds.   Verlag J. Peperny:Munich.   pp. 379-380.

Strassmann, J. E., D. C. Queller and C. R. Hughes.   1988. Predation and the evolution of sociality in the paper wasp, *Polistes bellicosus*. *Ecology*   69:1497-1505.

Stubblefield, J. W., and E. L. Charnov.   1986. Some conceptual issues in the origin of eusociality. *Heredity*   57:181-187.

Sullivan, J. D., and J. E. Strassmann.   1984.   Physical variability among nest foundresses in the polygynous social wasp, *Polistes annularis*. *Behav. Ecol. Sociobiol.*   15:249-256.

Thornhill, R. and J. Alcock.   1983.   *The Evolution of Insect Mating Systems*.   Harvard Univ. Press:Cambridge, Mass.

Trivers, R. L.   1974.   The evolution of reciprocal altruism. *Quart. Rev. Biol.*   46:35-39, 45-47.

Trivers, R. L. and H. Hare.   1976.   Haplodiploidy and the evolution of social insects. *Science*   191:249-263.

Vehrencamp, S. L. 1979. The roles of individual, kin and group selection in the evolution of sociality.   In *Handbook of Behavioral Neurobiology, vol. 3. Social Behavior and Communication*, P. Marler and J. Vandenbergh, eds.   Plenum:New York   pp. 351-394.

Ward, P. S.   1983.   Genetic relatedness and colony organization in a species complex of ponerine ants.   *Behav. Ecol. Sociobiol.* 12:285-299.

Ward, P. S., and R. W. Taylor.   1981.   Allozyme variation, colony structure and genetic relatedness in the primitive ant, *Nothomyrmecia macrops* Clark (Hymenoptera: Formicidae). *J. Aust. Entomol. Soc.* 20:177-183.

West-Eberhard, M. J. 1969. The social biology of polistine wasps. *Misc. Publ. of the Museum of Zool, U. of Michigan* 140: 1-101.

West-Eberhard, M. J. 1975. The evolution of social behavior by kin selection. *Quart. Rev. Biol.* 50:1-33.

West-Eberhard, M. J. 1978. Polygyny and the evolution of social behavior of wasps. *J. Kansas Entomol. Soc.* 51:832-856.

Willham, R. L. 1963. The covariance between relatives for characters composed of components contributed by related individuals. *Biometrics* 19:18-27.

Williams, G. C. and D. C. Williams. 1957. Natural selection of individually harmful social adaptations among sibs with special reference to social insects. *Evolution* 11:32-39.

Wilson, E. O. 1971. *The insect societies*. Belknap Press of Harvard Univ. Press:Cambridge, Mass.

Wilson, E. O. 1985. The sociogenesis of insects colonies *Science* 228:1489-1495.

Wilson, E. O. 1987. Causes of ecological success: the case of the ants. *J. Anim. Ecol.* 56:1-9.

Woolfenden, G. E. and J. W. Fitzpatrick. 1984. *The Florida scrub jay, demography of a cooperative-breeding bird.* Princeton University Press:Princeton, NJ.

CHAPTER 6

MEASURING INCLUSIVE FITNESS IN SOCIAL WASPS

David C. Queller and Joan E. Strassmann

Since Hamilton's (1964a 1964b) seminal work, students of the evolution of social behavior have become accustomed to thinking in terms of the inclusive fitness effects of behaviors. For example, the inclusive fitness effect of an altruistic behavior is $c + rb$ where c is the fitness cost to the altruist, b is the fitness gain of the beneficiary and r is their relatedness. The sign of the inclusive fitness effect determines the expected direction of selection. If the inclusive fitness effect of a behavior is positive, then selection should lead to an increase in the frequency of genes causing the behavior.

Inclusive fitness theory provides a simple way to think about the evolution of social interactions. While it is not as rigorous as the recursion models of population genetics, it is much easier to use. Any insights from inclusive fitness theory can be checked, and sometimes modified, through the use of more rigorous models, but inclusive fitness thinking helps modelers know what to model.

The inclusive fitness method of accounting has also played an important role in empirical studies. The alternative of simply measuring personal fitness can sometimes be used (Grafen 1982), but only when all individuals have the opportunity to express the behavior under study. Here its value goes beyond mere simplicity. There is a good reason why inclusive fitness has been the tool of choice for field studies of the evolution of kin relations. The empirical advantage of inclusive fitness arises when expression of the behavior is conditional on some special circumstance, so that not all individuals bearing the genes for the behavior will actually express it. Such cases are common. For

Department of Biology, Rice University, P. O. Box 1892, Houston TX 77251, USA example, among *Polistes* wasp foundresses, the most successful females are those who become queens of multiple-foundress associations, but obviously this alternative is not open to all foundresses. Those who are unable to attain this status must either nest alone or forego most of their personal reproduction by becoming a subordinate in a multiple-foundress association. Comparing the numbers of offspring of solitary nesters and subordinates is not sufficient for understanding how selection operates on this choice. To do so ignores the genes for the behavior that are transmitted by the offspring of the dominant queens. Just because the queens themselves do not express the choice of nesting alone or being a subordinate does not imply that they do not transmit the genes that affect this choice. The inclusive fitness solution is to assign those offspring of the queen that are in excess of what she would have had on her own and then devalue them by the relatedness of the subordinate to the queen. In effect, we use the observed behavior of some individuals, the subordinates, to predict the genes of other individuals who do not express the behavior, the queens. Some such procedure is necessary to make a complete accounting of the transmitted genes. Similar, and in some respects more sophisticated, methods are used by animal breeders (Lush 1947, Willham 1963, Falconer 1981). These could be used in some studies of social selection (Cheverud 1984), but in practice the simpler inclusive fitness method is generally preferred.

As inclusive fitness is the preferred method for dealing with this problem, it is curious that there have been so few attempts to measure inclusive fitness in social insects. More remarkable still is the fact that, after a small flush of studies conducted in the 1970's (Metcalf and Whitt 1977b, Gibo 1978, Noonan 1981, Strassmann 1981a) and later analyses of those studies (Grafen 1984, Queller and Strassmann 1988), there seem to have been no further serious attempts to measure inclusive fitness in social insects. While it is too soon to conclude that entomologists have given up on inclusive fitness, there is a distinct contrast with ornithologists, who continue to refine their inclusive fitness studies of helping in birds (e.g. Brown 1987, Woolfenden and Fitzpatrick 1984).

We can suggest two possible reasons for this difference. One might be a perception that we already understand the evolution of insect societies. This view may be held because of the widespread acceptance of Hamilton's (1964b) haplodiploid hypothesis which attributes the high incidence of eusociality in the haplodiploid Hymenoptera to the fact that super sisters are related by 3/4 instead of the usual 1/2. To the extent that this explanation is sufficient, there seems to be little left to explain. But recent reviews of the evolution of eusociality tend to downplay the special relatedness considerations that arise from haplodiploidy, pointing to costs and benefits as more important (Andersson 1984, Alexander and Noonan in press). It is clear that the question is not yet resolved and that field measurements of relatedness, costs, and benefits are important for its resolution.

The other reason for the rarity of studies of social insect inclusive fitness may be the difficulty of measuring important parameters. First, observation is difficult in most social insects because many important activities take place within an enclosed nest. It is no accident that *Polistes*, with its open nests, has been the favorite subject of inclusive fitness studies. Second, assigning paternity to individuals is very difficult because mated pairs do not remain together (except in termites). Finally, in many species, more than one female may oviposit in a single nest, making maternity difficult to assess as well. Some of these difficulties also apply to certain avian systems, but they arise more often and more consistently in social insects.

The solution to these difficulties is to devise methods that make inclusive fitness studies easier and more rigorous. The rest of the chapter is devoted to this task. We cannot address all the issues involved in measuring inclusive fitness so we focus on two recent efforts in our laboratory. One involves the measurement of relatedness and the other concerns a precondition necessary for accurate measurement of costs and benefits. We will use *Polistes* wasps as examples; however, the issues involved are of general interest.

ESTIMATING RELATEDNESS

The most common method of estimating relatedness uses pedigree path analysis (Cannings and Thomson 1981). A major advantage of this method is that it allows assignment of relatedness values between any pair of individuals whose pedigrees are known, but pedigrees are often very difficult to obtain for social insects. When pedigree data are inadequate, one can rely on assumptions, for example by assuming that all females mate only once with an unrelated male. It is better not to rely on untested assumptions. This uncertainty has led to increased reliance on protein electrophoresis data. The first studies used allozyme markers as an aid to pedigree construction (Metcalf and Whitt 1977a, Lester and Selander 1981, Pamilo 1982a, Page and Metcalf 1982). This can work quite well when there is only one kind of missing information. For example, if maternity can confidently be assigned through behavioral observations, then allozyme markers can then be used to estimate how many fathers each brood must have had, and relatedness of the brood can then be estimated. However, if there are too many unknowns, as is often the case, this method becomes difficult.

The solution to this problem is to jettison pedigree construction in favor of another approach. The alternative to the pedigree approach again makes use of allozyme data. The r in Hamilton's rule can be expressed as a genetic correlation or regression. Rather than going through an intermediate stage of pedigree construction, statistical estimates of these parameters are generated directly from allozyme data. The methodology for obtaining these estimates (Pamilo and Crozier 1982, Crozier *et al.* 1984, Pamilo 1984) has been employed in recent social insect studies in which pedigrees were hard to obtain (Craig and Crozier 1979, Pamilo and Varvio-Aho 1979, Ward and Taylor 1981, Pamilo 1982b, Pearson 1982, Ward 1983, Crozier *et al.* 1984, 1987, Ross and Fletcher 1985, Ross 1986, Schwartz 1986, 1987, Crozier and Pamilo 1986, Reilly 1987, Van der Have *et al.* 1988, see also Ross this volume, Kukuk this volume).

However, these statistical methods have at least two shortcomings. First, computer simulations show that the estimates of r are sometimes too low, particularly when there are data from relatively few social groups (Pamilo and Crozier 1982, Wilkinson and McCracken 1986). Second, these estimates yield only population averages, such as the average relatedness of foundresses to other foundresses on their nest. These population values are very useful, but it would also be helpful to have individual relatedness values. For example, we might wish to estimate the relatedness of foundresses on one nest only, or we may wish to compare their relatedness with that of foundresses on another nest. The methodology for attacking questions of this type has not been worked out.

Solving these problems (Queller and Goodnight, in press) should make allozyme studies considerably more valuable. A detailed treatment cannot be given here, but a summary will convey the main features of our approach. The quantity to be estimated is

$$r = \frac{\sum_{A} (p_y - \bar{p})}{\sum_{A} (p_x - \bar{p})} \qquad\qquad (1)$$

where the p's are Grafen's (1985) "p-values" which are usually interpreted as gene frequencies: $\bar{p}$ is the population frequency; p_x is its frequency in the actor, x; and p_y is its frequency in the actor's partners, y. The summations are taken over all altruistic acts, A, that are induced by the gene in question. This measure is closely related to the more familiar genetic regression, but it is slightly more general, that is, it makes Hamilton's rule work under a broader range of assumptions (Grafen 1985). The crucial difference is that our measure (1) takes deviations from the mean of the whole population while a regression measure takes deviations from the separate means of the actors and their partners ($r = \sum(p_y - \bar{p}_y)(p_x - \bar{p}_x) / \sum(p_x - \bar{p}_x)^2$).

The new measure (1), though less familiar than the regression coefficient, actually has a simpler interpretation in terms of selection for social behavior. If the p's represent frequencies of the gene for altruism, $p_y - \bar{p}$ represents the difference of a beneficiary from the population frequency and $p_x - \bar{p}$ represents the difference of an altruist from the same average. As selection affects a gene frequency to the extent that selected individuals differ from the population mean (there would be no genetic change if the selected group matched the population mean), the numerator and denominator of (1) must represent the relative importance of beneficiaries and altruists as vehicles for selection on the altruism gene. That is why this measure acts as a proper coefficient to scale the relative values of costs and benefits. Considered from the standpoint of a single copy of the altruism gene in an altruist, its identity with genes in the beneficiary is p_y, while its identity by chance alone is $\bar{p}$. Therefore $p_y - \bar{p}$ represents a measure of total identity minus identity by chance, in other words, identity by descent. Similarly $p_x - \bar{p}$ represents the allele's identity by descent with its own genotype (including itself, so $p_x \geq 1/2$ for diploids).

It is easy to see how neutral allozyme markers can be used to estimate r. Suppose we have available x altruists scored for k variable marker loci with a allele copies at each (here we mean allele copies, not allelo*morphs*, hence a always equals two for diploids). A reasonable estimator of (1) is

$$ r = \frac{\sum\limits_{x} \sum\limits_{k} \sum\limits_{a} (p_y - \bar{p}^*)}{\sum\limits_{x} \sum\limits_{k} \sum\limits_{a} (p_x - \bar{p}^*)} \tag{2} $$

In effect, we sum all identities by descent of altruist marker genes with beneficiary genes and divide by the sum of all identities by descent of altruist marker genes with genes in their own genotypes. (Note that to avoid clutter we have not indexed the p-values for alleles, but they do change according to which allele is currently being summed over). Often we are interested not so much in actual altruists as potential ones. If we let x be an index for potential altruists and y for potential beneficiaries, the formula (2) still applies.

The only complication concerns the estimate of the population frequency for any allele. Instead of $\bar{p}$, we have written it as $\bar{p}^*$ to indicate a necessary bias correction when a small number of altruists or potential altruists are sampled. Here $\bar{p}^*$ is the population mean frequency of the allele currently being summed over, but calculated after omission of the current altruist, x, and its beneficiaries, y. To understand the rationale for this, consider the effect of not excluding these individuals. In a large population these individuals' genes make a negligible contribution to the actual population gene frequency, but they would make a sizable contribution to an estimate calculated from a small sample. This presents no problem for an estimate of the population frequency itself, but it leads to biased estimates of p_x-$\bar{p}$ and p_y-$\bar{p}$. For example, since each x individual contributes an artificially large proportion to $\bar{p}$, the difference, p_x-$\bar{p}$, will be artificially reduced. The solution is to omit x (and its genetically similar relatives, y) from the estimate of $\overline{p}$, so that the difference, $p_x-\overline{p}$ *, is now taken using an estimate of the population frequency which, like the true population frequency, is essentially unaffected by individuals currently being considered. Failure to make such a correction causes the previously noted (Pamilo and Crozier 1982, Wilkinson and McCracken 1986) downward bias of the regression estimator. Simulations show that using $\overline{p}$ * does in fact correct this bias (Queller and Goodnight, in press).

The second useful feature of this method is that it provides a clear and simple way to estimate individual r values rather than just population averages. By individual r values we mean either the average relatedness between one altruist and its beneficiaries or the relatedness within one particular social group. For individual estimates, the regression measure has a zero denominator and is, therefore, undefined. One can visualize this by recognizing that there is no unique regression line that can be fit to a single point. In contrast, individual r values obtained directly from formula (1) are defined. Summations are performed for the single altruist in question (or for all the members of the single group of interest) instead of the entire sample. It is important to remember, however, that the $\overline{p}$ * values must still be calculated using *other* individuals. What is important is how much the altruist and its beneficiary differ from the population mean, so it is still necessary to gather genetic information from more than just the individuals of interest.

Because the estimates of individual r's are highly dependent on the genotypes of just a few individuals, they are expected to be much more variable than estimates of average population relatedness. This means that they will be fairly unreliable unless there is a great deal of genetic information available. Although each individual value may be quite variable, they can be used together to test interesting hypotheses. Suppose, for example, we want to know if foundresses who associate before March 25 are more closely related than those that first associate after this date. We can calculate average r's for each group, but we need to know something about the dispersion of values to test for statistical significance. One simple way to determine the dispersion is to estimate r separately for each foundress association and use a non-parametric Mann-Whitney U test to compare those formed before and after March 25. Simulations have shown that this kind of procedure can be quite successful at detecting differences in relatedness, even when the individual estimates are quite unreliable (Queller and Goodnight, in press).

We have not yet applied these methods to many examples of real data, but the following example gives another indication of the kinds of hypotheses that can be investigated. At the end of a reproductive season, *Polistes* colonies rear reproductive males and females. The females will overwinter and begin new nests the following spring, often in association with natal nestmates. It is, therefore, of interest to know the relatedness of reproductive females on their natal nests, and whether relatedness differs on different kinds of nests. For example, one might hypothesize that relatedness might be lower on natal nests that produce many reproductives because the queen might have had more trouble monopolizing reproduction on larger and more productive nests. FIGURE 1 shows data from a study we have conducted with Colin R. Hughes on *Polistes instabilis* from McAllen, Texas that can be used to test this hypothesis. Two polymorphic loci were scored for all females taken from 41 fall nests. The plot of colony relatedness values versus number of females shows that the hypothesis is not supported. There is a trend in the opposite direction: relatedness tends to increase with higher productivity (though the Spearman's rank correlation coefficient of .247 is not significant).

COSTS AND BENEFITS

Fitness is always a difficult parameter to define and to measure. Measurement of costs and benefits, which are components of fitness, should share the same difficulties. However, it is not these difficulties that we want to address here. Approximate measures, such as counting offspring at a certain stage, are as acceptable in social behavior as they are in other fields. But there are two problems that are peculiar to inclusive fitness studies, and we focus on these.

The first is an accounting problem that we treat very briefly as it has been discussed by Grafen (1982). While traditional personal fitness models allocate all offspring to their parents' fitness, inclusive fitness models allocate offspring to the fitness of the individual that is responsible for them. When a helper increases its sister's reproduction, inclusive fitness models augment the fitness of the helper. However, care must be taken not to allocate the same fitness units to more than one

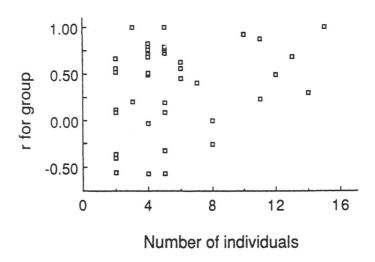

FIGURE 1. Relatedness of *Polistes instabilis* females as a function of group size. Each value represents relatedness within one colony collected in the fall just prior to overwintering.

individual. In the example above, the offspring cannot also be added to the sister's fitness. Grafen (1982) noted that inclusive fitness has often been defined in ways that incorporate such double counting, but that this has less often been a problem in actual measurements of inclusive fitness. Measurements are usually done using Hamilton's rule, and the structure of this rule forces us to think in a way that avoids double counting.

However, there remains a second general problem with trying to estimate costs and benefits. Costs and benefits are defined as fitness effects of a behavior (Hamilton 1964). That is, they are *differences* due to the behavior; they are each defined as a fitness actually achieved minus the fitness that would have pertained if the behavior had not occurred. The special problem with measuring inclusive fitness lies with this second part. How do we estimate the number of offspring that would have been produced in the absence of the behavior?

As an example, consider the choice facing an overwintered *Polistes* female who would not, for whatever reason, be able to attain dominant queen status in a multiple-foundress association. She can accept the status of a subordinate and help to rear the queen's offspring or she can establish a nest on her own. This is the choice investigated in most inclusive fitness studies of *Polistes* (Metcalf and Whitt 1977b, Gibo 1978, Noonan 1981, Strassmann

1981a, Grafen 1984, Queller and Strassmann 1988). What are the costs and benefits of joining as a subordinate? The cost is a subordinate's personal fitness minus the fitness she gave up by not nesting alone. Similarly, the benefit is the fitness that the subordinate adds to her relative. Estimating this requires knowing what the relative's fitness would have been without the subordinate's help. Neither the cost nor the benefit is directly measurable. In the inclusive fitness studies cited above, the way out of this dilemma was to use the fitness of solitary foundresses to represent both how well a subordinate would have done if she had nested alone and how well a queen would have done without the help of any helpers. As the authors of these studies recognized, this requires an assumption that other things are equal. This is a familiar assumption in studies of adaptation, and it often works well, but there is a special reason to distrust it in this case. Note that there are two assumptions required: that subordinates and solitary foundress would be equally capable at nesting alone, and that queens and solitary females would be equally capable. Each assumption seems plausible enough by itself, but together they imply that queens and subordinates are equally capable at nesting alone. Yet queens and subordinates are unequal in some respects; otherwise the queen would not be able to dominate her subordinates. Queens and subordinates, depending on the species and population, differ in factors such as size, ovarian development, hormone levels, and fat content (Pardi 1946 1948, West-Eberhard 1969, Turillazzi and Pardi 1977, Röseler *et al.* 1980 1984, Dropkin and Gamboa 1981, Turillazzi *et al.* 1982, Sullivan and Strassmann 1984). While some of these differences could be partly a result of dominance interactions, it is clear that queens and subordinates are not identical. Moreover, it has been argued that dominance should be expected to reflect reproductive ability (West-Eberhard 1978). To the extent that this is true, the standard assumptions of inclusive fitness analyses are false, and the inclusive fitness estimates are in error.

It is possible that, in spite of the behavioral, morphological, and physiological differences between queens and subordinates, they would still be equally capable of nesting alone. But given the importance of the assumption, and the distinct possibility that it is false, it needs to be tested. We have attempted to do so during the course of a study of *Polistes bellicosus* at Brazos Bend State Park near Houston, Texas. Full details of this study will be presented elsewhere (Queller and Strassmann, in preparation). Here we focus on the single issue of the reproductive equivalence of queens and subordinates.

We began observing *Polistes bellicosus* colonies from the time of their founding in late March, 1985 (Strassmann *et al.* 1987). Nests were usually begun by a single female but were often subsequently joined by others. On April 6 we performed our experimental manipulation. At this stage most nests had early instar larvae and three quarters of the colonies had more than one female, all of which had been marked with enamel. Multiple-female colonies were randomly assigned to three treatment groups. We removed all but the largest female from 1/3, all but the smallest female from 1/3 and left the rest untreated. A fourth group, which we did not alter, consisted of the colonies that only had one female to start with. Our main interest here is in the three treatments in which a single female is left to maintain the nest.

If the largest female in a colony is usually the queen and the smallest female is usually a subordinate, then our two experimentally manipulated groups would consist primarily of queens forced to nest alone and subordinates forced to nest alone. Size is a well-known correlate of dominance in *Polistes* foundresses (Turillazzi and Pardi 1977, Noonan 1981, Sullivan and Strassmann 1984) and *Polistes bellicosus* is no exception. Due to the large numbers of colonies in our experiment, we did not assess behavioral dominance directly. But queenship could be inferred from dissections of wasps from collected colonies. In ten out of fourteen complete colonies collected, the largest female was judged to be the queen on the basis of clearly having the best-developed ovaries. On only one was the smallest female the queen. This trend was confirmed by dissections of females from the colonies that were partially collected during the experimental manipulation. When we collected all but the largest female from some colonies, we obtained a sample of smallest females for dissection, and when we collected all but the smallest females we obtained a sample of largest females. Dissections showed the largest females to have significantly better developed ovaries than the smallest females for every measure taken (p<0.05, Mann Whitney U tests for: number of layable eggs, number of oocytes per ovariole, number of oocytes with yolk in the best-developed ovariole, lengths of the largest and the second largest oocytes in that ovariole). We conclude that largest females are usually queens and smallest females almost never are. Therefore, even

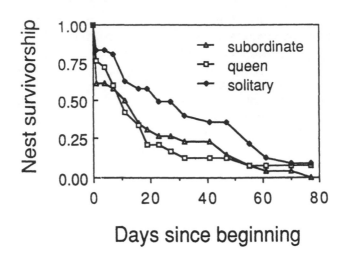

FIGURE 2. Survivorship of colonies tended by three categories of single females after experimental manipulation on April 6. Subordinates and queens are individuals who had those roles on multifemale associations prior to the removal of their nestmates on April 6. Solitary females are those who were already nesting alone prior to April 6.

though there are likely to be a few exceptions, we will call the largest females queens and the smallest females subordinates.

Did queens fare better than subordinates when forced to nest alone? The answer was no. We censused nests repeatedly until June 21, by which time nearly all had failed. Failures were due either to death of all adults, desertion by all adults, or predation on the nest comb (after failure due to the latter two causes, females may begin or join another nest, but our measures of success pertain only to the original nests). Queens nesting alone did not maintain their colonies for a longer time than subordinates nesting alone (FIGURE 2). Nor were there any significant differences, at any census date after the experimental manipulation, in any measures of brood rearing success (number of cells, eggs, larvae, or pupae). In all respects, queen colonies and subordinate colonies fared equally well.

Actually, it would be more accurate to say that the two groups fared equally poorly, and this shift in emphasis may have some significance in the interpretation of our results. The fact that queens and subordinates fared poorly when nesting alone is shown by comparison with the colony survivorship of natural colonies of solitary foundresses. The females who chose to nest on their own maintained their colonies for significantly longer than those that we forced to nest alone (FIGURE 2). The explanation would seem to lie in the fact that, prior to our manipulation, the experimental colonies had a brief history as multiple-foundress colonies. Because of this, the experimental nests were larger than the control solitary nests (FIGURE 3). By removing foundresses from a nest, we may have overburdened the remaining foundress with a larger brood than she would have chosen to raise if she had been a solitary foundress from the beginning. Her response could be to increase foraging, thereby increasing risk of mortality, or she might opt to desert the nest and go elsewhere. We cannot distinguish between these alternatives, but either could account for the decreased survivorship of experimental colonies.

Our conclusions, then, are less straightforward than we might have hoped, but still revealing. A simple comparison of queens and subordinates suggests that they are reproductively equivalent when nesting alone, so that the equivalence assumption implicit in the use of Hamilton's rule seems justified. But the experimental reduction in foundress numbers may have stressed the remaining females, and this conclusion could be challenged on two grounds. First, it might be argued that our experimental treatment was unnatural and that, therefore, the results could not be taken as indicative of the actual abilities of queens and subordinates under normal circumstances. But in fact the stress engendered by the experiment is not unnatural. Mortality of *Polistes* foundresses is high and reduction of multiple-foundress colonies to a single female is a common event. Indeed, we found that the death or disappearance of all foundresses was the most common cause of colony failure prior to the emergence of workers.

It could also be argued that we did not test all of the abilities required to nest alone. Specifically, if our queens and subordinates were overburdened with brood after the experimental manipulation, one would not expect them to lay additional eggs, so our experiment would not test their ability to perform this task. The data in FIGURE 3 confirm that former queens and subordinates did not add new cells to their nests after the experimental removal of other females. However, FIGURE 3 also suggests that this is the normal pattern for solitary females during this period since the unmanipulated single-foundress nests did not grow either. Apparently solitary females begin a cohort of brood and then concentrate on rearing that brood before laying

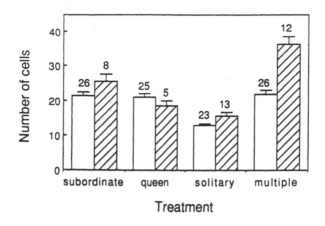

FIGURE 3. Average numbers of cells in nests on the day of the experiment, April 6 (open bars), and 22 days later (hatched bars). Categories of nests are as in FIGURE 2, with the addition of "multiples", which are colonies with more than one foundress before April 6 that did not have any females removed.

additional eggs. In contrast, multiple-foundress nests continue to grow throughout this period.

We can conclude, then, that our experiment lends some support to the assumption that queens and subordinates are reproductively equivalent in their brood-rearing abilities. While we did not test relative egg-laying abilities, this is not very relevant at the time of year when our study was conducted anyway.

DISCUSSION AND CONCLUSIONS

We have not discussed all the issues relevant to conducting accurate and meaningful inclusive fitness studies in wasps like *Polistes*. But we have removed two serious obstacles. On the one hand, we have developed a statistical estimator of relatedness that is both unbiased and flexible enough to estimate individual relatedness values. On the other, we tested a critical assumption required for rigorous measurement of costs and benefits and we have failed to find anything wrong with it. Of course, these two advances do not eliminate the hard work that will be necessary to measure relatedness, costs, and benefits. But they do suggest that such efforts will yield valid information. Perhaps the time is ripe for a renewal of interest in inclusive fitness studies of social insects.

Even measurements of relatedness alone, without costs and benefits, can be valuable in answering important questions. What benefit/cost ratio is required for sociality to pay, or in other words, how efficient do groups have to be? How do individuals use their information about relatedness in making behavioral choices? To what extent do intraspecific differences in *r* predict differences between species in the extent of altruism? With Colin Hughes, we are currently trying to answer such questions for several Texas *Polistes* species. Beyond a few temperate *Polistes* and vespines (Ross 1986), almost nothing is known of relatedness in social wasps. Some of the tropical polybiine wasps offer some especially intriguing opportunities for relatedness studies. Many of these species are characterized by multiple queens, suggesting that relatedness may be quite low. In addition, most of these species reproduce by swarming and it would be very interesting to know what role relatedness plays in behavioral decisions during the colony fissioning stage.

Similarly, much remains to be learned about costs and benefits. The testing of alternatives focusing on costs and benefits is still in the early stages (see Strassmann and Queller, chapter 5). Particularly needed are demonstrations of how an altruist can give more, in terms of fitness, to its relative than it could give to itself. *Polistes* is the best-studied group, and some important benefits of helping have been proposed (West-Eberhard 1969, Metcalf and Whitt 1977b, Gibo 1978, Gamboa 1978 1980, Noonan 1981, Strassmann 1981b, Strassmann *et al.* 1988), but it is probably fair to say that no general picture has yet emerged. Once again, there has been far too little work on the polybiines, with the exception of *Mischocyttarus*, which is behaviorally rather like *Polistes* (Jeanne 1972, Litte 1977 1979). Studies of the costs and benefits of sociality in other polybiines could be particularly illuminating with respect to the evolution of castes because this group contains species with different grades of morphological caste differentiation, including none at all.

Acknowledgements

We thank Carlos Solís for field assistance and for doing the figures, Keith Goodnight for writing the computer program for estimating relatedness, and Colin Hughes for help with field work and for generating the electrophoresis data of FIGURE 1. Part of the work reported here was conducted at Brazos Bend State Park under Texas Parks and Wildlife permit number 16-85, and we thank D. Riskind, L.

Fowler, and C. VanBaarle for their cooperation. This work was partially supported by NSF grant BSR-8695026.

LITERATURE CITED

Alexander, R. D. and K. M. Noonan. In press. The evolution of eusociality. In *The natural history and social behavior of naked mole rats*, P. W. Sherman, J. Jarvis and R. D. Alexander, eds.

Andersson, M. 1984. The evolution of eusociality. *Ann. Rev. Ecol. Syst.* 15:165-189.

Brown, J. L. 1987. *Helping and communal breeding in birds: ecology and evolution.* Princeton University Press:Princeton, NJ.

Cannings, C. and E. A. Thomson. 1981. *Genealogical and genetic structure.* Cambridge University Press:Cambridge, U. K.

Cheverud, J. 1984. Evolution by kin selection: a quantitative genetic model illustrated by maternal performance in mice. *Evolution* 38:766-777.

Craig, R. and R. H. Crozier. 1979. Relatedness in the polygynous ant *Myrmecia pilosula. Evolution* 33:335-341.

Crozier, R. H. and P. Pamilo. 1986. Relatedness within and between colonies of a queenless ant species of the genus *Rhytidoponera* (Hymenoptera: Formicidae). *Entomol. Gener.* 11:113-117.

Crozier, R. H., P. Pamilo and Y. C. Crozier. 1984. Relatedness and microgeographic genetic variation in *Rhytidoponera mayri*, an Australian arid zone ant. *Behav. Ecol. Sociobiol.* 15:143-150.

Crozier, R. H., B. H. Smith and Y. C. Crozier. 1987. Relatedness and population structure in the primitively eusocial bee *Lasioglossum zephyrum* (Hymenoptera: Halictidae) in Kansas. *Evolution* 41:902-910.

Dropkin, J. A. and G. J. Gamboa. 1981. Physical comparisons of foundresses of the paper wasp, *Polistes metricus* (Hymenoptera: Vespidae). *Can. Entomol.* 113:457-461.

Falconer, D. S. 1981. *Introduction to quantitative genetics.* 2nd ed. Longman:London.

Gamboa, G. J. 1978. Intraspecific defense: advantage of social cooperation among paper wasp foundresses. *Science* 199:1463-1465.

Gamboa, G. J. 1980. Comparative timing of brood development between multiple and single-foundress colonies of the paper wasp, *Polistes metricus. Ecol. Entomol.* 5:221-225.

Gibo, D. L. 1978. The selective advantage of foundress associations in *Polistes fuscatus* (Hymenoptera: Vespidae): a field study of the effects of predation on productivity. *Can. Entomol.* 110:519-540.

Grafen, A. 1982. How not to measure inclusive fitness. *Nature* 298:425-426.

Grafen, A. 1984. Natural selection, kin selection and group selection. In *Behavioural ecology: an evolutionary approach,* J. R. Krebs and N. B. Davies, eds. Sinauer:Sunderland, Mass. pp. 62-84.

Lush, J. L. 1947. Family merit and individual merit as bases for selection. *Am. Nat.* 81:241-261.

Hamilton, W. D. 1964a. The genetical evolution of social behavior. I. *J. Theoret. Biol.* 7:1-16.

Hamilton, W. D. 1964b. The genetical evolution of social behavior. II. *J. Theoret. Biol.* 7:17-52.

Jeanne, R. L. 1972. Social biology of the neotropical wasp *Mischocyttarus drewseni. Bull. Mus. Comp. Zool. Harvard* 144:63-150.

Lester, L. J. and R. K. Selander. 1981. Genetic relatedness and the social organization of *Polistes* colonies. *Am. Nat.* 117:147-176.

Litte, M. 1977. Behavioral ecology of the social wasp, *Mischocyttarus mexicanus. Behav. Ecol. Sociobiol.* 2:229-246.

Litte, M. 1979. *Mischocyttarus flavitarsus* in Arizona: social and nesting biology of a polistine wasp. *Z. Tierpsychol.* 50:282-312.

Metcalf, R. A. and G. S. Whitt. 1977a. Intra-nest relatedness in the social wasp *Polistes metricus:* a genetic analysis. *Behav. Ecol. Sociobiol.* 2:339-351.

Metcalf, R. A. and G. S. Whitt. 1977b. Relative inclusive fitness in the social wasp *Polistes metricus. Behav. Ecol. Sociobiol.* 2:353-360.

Noonan, K. M. 1981. Individual strategies of inclusive-fitness-maximizing in *Polistes fuscatus* foundresses. In *Natural selection and social behavior: recent research and new theory,* R. D. Alexander and D. W. Tinkle, eds. Chiron Press:New York pp. 18-44.

Page, R. E. and R. A. Metcalf. 1982. Multiple mating, sperm utilization, and social evolution. *Am. Nat.* 119:263-281.

Pamilo, P. 1982a. Multiple mating in *Formica* ants. *Hereditas* 97:37-45.

Pamilo, P. 1982b. Genetic population structure in polygynous *Formica* ants. *Heredity* 48:95-106.

Pamilo, P. 1984. Genetic correlation and regression in social groups: multiple alleles, multiple loci, and subdivided populations. *Genetics* 107:307-320.

Pamilo, P. and R. H. Crozier. 1982. Measuring genetic relatedness in natural populations: methodology. *Theoret. Popul. Biol.* 21:171-193.

Pamilo, P. and S.-L. Varvio-Aho. 1979. Genetic structure of nests of the ant *Formica sanguinea. Behav. Ecol. Sociobiol.* 6:91-98.

Pardi, L. 1946. Richerchi sui polistini. V. La poliginia iniziale in *Polistes gallicus* L. *Boll. Ist. Ent. Univ. Bologna* 15:25-84.

Pardi, L. 1948. Dominance order in *Polistes* wasps. *Physiol. Zool.* 21:1-13.

Pearson, B. 1982. Relatedness of normal queens (macrogynes) in nests of the polygynous ant *Myrmica rubra* Latreille. *Evolution* 36:107-112.

Queller, D. C. and K. F. Goodnight. in press. Estimation of genetic relatedness using allozyme data. *Evolution.*

Queller, D. C. and J. E. Strassmann. 1988. Reproductive success and group nesting in the paper wasp, *Polistes annularis.* In *Reproductive success: studies of individual variation in contrasting systems.* T. H. Clutton-Brock, ed. University of Chicago Press:Chicago. pp. 76-96.

Reilly, L. M. 1987. Measurements of inbreeding and average relatedness in a termite population. *Am. Nat.* 130:339-349.

Röseler P.-F., I. Röseler and A. Strambi. 1980. The activity of corpora allata in dominant and subordinated females of the wasp *Polistes gallicus. Insectes Sociaux* 27:97-107.

Röseler P.-F., I. Röseler, A. Strambi and R. Augier. 1984. Influence of insect hormones on the establishment of dominance hierarchies among foundresses of the paper wasp, *Polistes gallicus. Behav. Ecol. Sociobiol.* 15:133-142.

Ross, K. G. 1986. Kin selection and the problem of sperm utilization in the social insects. *Nature* 323:798-800.

Ross, K. G. and D. J. C. Fletcher. 1985. Comparative study of genetic and social structure in two forms of the fire ant *Solenopsis invicta* (Hymenoptera: Formicidae). *Behav. Ecol. Sociobiol.* 17:349-356.

Schwartz, M. P. 1986. Persistent multi-female nests in an Australian allodapine bee, *Exoneura bicolor* (Hymenoptera: Anthophoridae). *Insectes Sociaux* 33:258-277.

Schwartz, M. P. 1987. Intra-colony relatedness and sociality in the allodapine bee *Exoneura bicolor. Behav. Ecol. Sociobiol.* 21:387-392.

Strassmann, J. E. 1981a. Wasp reproduction and kin selection: reproductive competition and dominance hierarchies among *Polistes annularis* foundresses. *Fla. Entomol.* 64:74-88.

Strassmann, J. E. 1981b. Parasitoids, predators and groups size in the paper wasp, *Polistes exclamans. Ecology* 62:1225-1233.

Strassmann, J. E., D. C. Queller and C. R. Hughes. 1987. Constraints on independent nesting by *Polistes* foundresses in Texas. In *Chemistry and biology of social insects,* J. Eder and H. Rembold, eds. . Verlag J. Peperny:München. pp. 379-380.

Strassmann, J. E., D. C. Queller and C. R. Hughes. 1988. Predation and the evolution of sociality in the paper wasp, *Polistes bellicosus.* Ecology 62:1225-1233.

Sullivan, J. D. and J. E. Strassmann. 1984. Physical variability among nest foundresses in the polygynous social wasp, *Polistes annularis. Behav. Ecol. Sociobiol.* 15:249-256.

Turillazzi, S. and L. Pardi. 1977. Body size and hierarchy in polygynic nests of *Polistes gallicus* (L.) (Hymenoptera Vespidae). *Monitore Zool. Ital. (N.S.)* 11:101-112.

Turillazzi, S., M. T. Marino Piccioli, L. Hervatin and L. Pardi. 1982. Reproductive capacity of single foundresses and associated foundress females of *Polistes gallicus* (Hymenoptera Vespidae). *Monitore Zool. Ital. (N.S.)* 16:75-88.

Van der Have, T. M., J. J. Boomsma and S. B. J. Menken. 1988. Sex investment ratios and relatedness in the monogynous ant *Lasius niger* (L.). *Evolution* 42:160-170.

Ward, P. S. 1983. Genetic relatedness and colony organization in a species complex of ponerine ants. *Behav. Ecol. Sociobiol.* 12:285-299.

Ward, P. S. and R. W. Taylor. 1981. Allozyme variation, colony structure and genetic relatedness in the primitive ant *Nothomyrmecia macrops* Clark (Hymenoptera: Formicidae). *J. Aust. Entomol. Soc.* 20:177-183.

West-Eberhard, M. J. 1969. The social biology of polistine wasps. *Misc. Publ. Museum Zool. Univ. Mich.* 140:1-101.

West-Eberhard, M. J. 1978. Polygyny and the evolution of social behavior of wasps. *J. Kansas Entomol. Soc.* 51:832-856.

Wilkinson, G. S. and G. F. McCracken. 1986. On estimating genetic relatedness using genetic markers. *Evolution* 39:1169-1174.

Willham, R. L. 1963. The covariance between relatives for characters composed of components contributed by related individuals. *Biometrics* 19:18-27.

Woolfenden, G. E. and J. W. Fitzpatrick. 1984. *The Florida scrub jay: demography of a cooperatively breeding bird.* Princeton University Press:Princeton, NJ.

CHAPTER 7

GENETIC AND SOCIAL CHANGES ASSOCIATED WITH ANT SPECIATION

Philip S. Ward

The salient features of eusocial behavior--overlapping generations, cooperative brood care, and (most critically) reproductive division of labor--appear to be part of the basic ground plan of the family Formicidae (Brothers 1975, Wilson 1987). Nevertheless, ant species vary considerably with respect to certain individual- and colony-level social traits, such as the reproductive proclivity of workers, mode of colony foundation, number of functional queens in a colony, division of labor among workers, extent of internest cooperation, and degree of dependence on alien colonies for worker labor (Wilson 1971, Hölldobler and Wilson 1977, Crozier 1979, Oster and Wilson 1978, Brian 1983).

In this paper I explore the question of whether the origin of certain types of social behavior in ants is likely to be closely linked to the development of reproductive isolation between populations. In other words, to what extent can the diversity of social organization exhibited by ants be attributed to conditions associated with speciation, as opposed to a more general process of phyletic (within-lineage) change? This question is worthwhile, not only because it bears upon the general issue of whether phenotypic change is concentrated around speciation events (Gould and Eldredge 1977, Charlesworth *et al.* 1982, Futuyma 1987), but also because the suggestion has frequently arisen in the ant literature of a possible connection between speciation and social change, particularly with regard to the development of polygyny and social parasitism (Wilson 1971, Crozier 1977, Elmes 1978, Brian 1983, Ross and Fletcher 1985, Ward 1985, Buschinger

Department of Entomology, University of California, Davis, CA, USA, 95616

1986, Ross *et al*. 1987b, West-Eberhard 1986 1987). It has been asserted for example, that pairs or sets of sibling species, with contrasting patterns of social organization, are common among the ants, with the implication that the differences arose during recent speciation events. Moreover theoretical considerations suggest that ant species, as *haplodiploid* organisms, might diverge more rapidly than diploid species (Hartl 1972, Avery 1984), at least for traits affected by favorable recessive alleles and chromosomal rearrangements (Charlesworth *et al*. 1987).

In attempting to assess these claims empirically, it becomes obvious that our knowledge of the speciation process in ants is quite deficient. Thus, my conclusions are necessarily tentative, and based primarily on weak inferences from patterns of species distributions and from comparisons of phenotypic differences among related species. The examples to be discussed here are taken largely from two groups of ants (the Indo-Australian genus *Rhytidoponera* and the New World genus *Pseudomyrmex*) in which I have studied systematic relationships and colony structure, but I also review the available evidence from other taxa. In many groups of ants the species-level taxonomic relationships have not been fully clarified and this limits our ability to make productive comparisons.

Consideration is first given to possible modes of speciation in ants, with an emphasis on the geographic relations of differentiating populations and on the kinds of genetic changes which might be expected to occur. After this background I will examine intra- and inter-specific patterns of variation in the social organization of ant colonies.

MODES OF SPECIATION IN ANTS

It is easier to consider the geographic aspects of ant speciation than the types of genetic changes involved because of the paucity of data on genetic differences between closely related ant species and on the kinds and intensity of reproductive isolating mechanisms. Here, to discern something about the nature of speciation in this group of organisms, I rely primarily on information about (1) geographical variation within species, (2) geographical distributions of closely related taxa, and (3) morphological differences among species.

Geographical Variation in Ants

There are numerous cases of apparently polytypic ant species showing extensive geographical variation in morphology. A few of the many available examples include Nearctic and Palearctic species in the genera *Lasius* (Wilson 1955), *Formica* (Francoeur 1973), and *Camponotus* (Creighton 1950); various ponerine ants found in Melanesia and Australia (Brown 1958, Wilson 1958); New World army ants in the genus *Eciton* (Borgmeier 1955); and African species of the genus *Tetramorium* (Bolton 1980). This polytypy is reflected in the plethora of subspecific and varietal names appearing in the older literature on ant taxonomy. A favorite example of mine is *Camponotus maculatus*, for which Emery (1925) lists 42 infraspecific names, many from the same geographical region!

Recent electrophoretic and cytological studies demonstrate that sibling species are common in ants (Crosland and Crozier 1986, Crozier 1977b 1981, Halliday 1981, Imai *et al.* 1977, Imai and Taylor 1986, Ward 1980a 1980b 1984). Thus, many polytypic ant "species" may in fact comprise several distinct biological species. This particularly applies to sympatric "subspecies" and "varieties". Nevertheless, when detailed species-level studies have been carried out, some of the lowest ranking units remain highly variable.

Several species groups of the Indo-Australian ant genus *Rhytidoponera* have been revised using a combination of electrophoretic and morphological characters (Ward 1980a, 1984). Multiple (sibling) species consistently appeared in cases in which previously only one was suspected; and the taxa thus resolved behave as good biological species in any given locality. Yet some *Rhytidoponera* species remain quite geographically variable in morphology and in allozymes. For two species in the *R. impressa* group, the interpopulation variation in allozyme gene frequencies as measured by Wright's (1965) F_{ST} statistic was higher than that reported for most outbreeding diploid species (F_{ST} values of .294 and .380, compared to .009-.162 across a range of diploid species; Ward 1980b). Ross *et al.* (1987b) reported low F_{ST} values for United States populations of the introduced fire ant, *Solenopsis invicta*, but this is not surprising in view of the recent origin of these populations. It would be interesting to have comparable statistics for *S. invicta* (or other members of *S. saevissima* complex) in their native ranges in

125

South America. Pamilo (1983) found moderate differentiation (F_{ST} .074-.090) among *Formica* populations occupying islands in southern Finland. These and most other electrophoretic studies have not revealed significant inbreeding coefficients (F_{IS}) within populations. There has been no compelling evidence of partially isolated sympatric entities. One possible exception concerns the sympatric "macrogyne" and "microgyne" forms of *Myrmica rubra* for which Pearson and Child (1980) (see also Pearson 1981) reported different allozyme frequencies.

The frequent occurrence of polytypy and sibling species in ants is consistent with the hypothesis that conspecific populations are rather prone to undergo divergence. The geographical variation exhibited by widespread taxa, and the generally discrete nature of ant species on a microgeographic scale, suggests that population differentiation is primarily an allopatric process. Nevertheless the large number of taxonomically unresolved complexes in the Formicidae leaves open the possibility that some of these involve more complicated kinds of reproductive relationships between sympatric populations. Such problematic cases include the macrogyne and microgyne forms of certain *Myrmica* species (Brian 1983); various taxa in the *Formica rufa* group (Vespsäläinen and Pisarski 1981); and sympatric "color morphs" of *Pseudomyrmex* (Ward 1985) and *Myrmecia* (Browning unpublished). Hybridization between ant taxa has been invoked to explain certain cases of taxonomic intermediacy (e.g. Wing 1968, Ward 1980a), but the only well documented case of a hybrid zone in ants is that between two introduced species of fire ants, *Solenopsis invicta* and *S. richteri,* in southeastern United States (Ross *et al.* 1987a).

Geographical Distribution of Closely Related Taxa

The lack of evidence for a role of polyploidy in ant evolution (Imai *et al.* 1977) suggests that speciation is a gradual process in ants. The study of ant speciation thus necessarily includes historical analysis. Yet we lack historical records of ant populations over the time period that speciation is likely to have occurred and the present-day distribution patterns of species may not reflect their geographical disposition during speciation. It would be desirable to examine species which have only recently separated, whose phylogenetic affinities are understood, and whose dispersal abilities are limited. Potentially useful for these purposes are several species-rich ant taxa whose reproductive females are entirely wingless, as the history of such populations should be less obscured by dispersal than those of highly vagile ants. Examples include the Old World and New World army ants (Dorylinae and Ecitoninae, respectively) (Gotwald 1982); most members of the speciose ponerine genera *Leptogenys* and *Rhytidoponera* (Wheeler 1923, Brown 1958); the dolichoderine genus *Leptomyrmex* (Wheeler 1934); and numerous smaller genera or species groups within genera (e.g. Wheeler 1917, Brown 1975, Bolton 1986). Note that in these cases gene flow itself is not necessarily curtailed--the males are always fully winged and capable of dispersal--but the movement of populations is.

Phylogenetic and geographical data are available for some species groups within the Indo-Australian genus *Rhytidoponera* (Ward 1980a 1980b 1984 1985b, Crozier *et al.* 1986). All eighteen species of *Rhytidoponera* on the island of New Caledonia lack winged female reproductives and in this fauna most pairs or triplets of closely related species are distributed allopatrically or show slight overlap (Ward 1984 1985b). More distantly related species are often broadly sympatric.

Among the New Caledonian *Rhytidoponera,* I came across several wide-ranging taxa composed of variably (and sometimes strongly) differentiated allopatric populations for which only arbitrary resolutions of species status were possible (Ward 1984). This problem of determining conspecificity among allopatric populations, called by Brown (1957) the uncertainty principle in systematics, is particularly acute among some species groups of *Rhytidoponera* that occur in drier habitats on the Australian mainland (Ward unpublished), and strongly suggests an allopatric (and non-punctuated) mode of speciation in these ants.

The *Rhytidoponera impressa* group is distributed along the east coast of Australia as a north-south series of partially overlapping populations (Ward 1980a). These ants are confined to wet forests (primarily rainforest) which must have undergone successive expansion and contraction during past climatic fluctuations. Some of the five species in this group (notably *R. chalybaea* and *R. impressa*) are quite heterogeneous and contain divergent allopatric populations which could be viewed as incipient species. A phylogenetic analysis of these species, combined with a consideration of their geographical distributions, yielded a pattern (FIGURE 1) which is consistent with a history of allopatric speciation, followed by partial sympatry. *R. enigmatica*

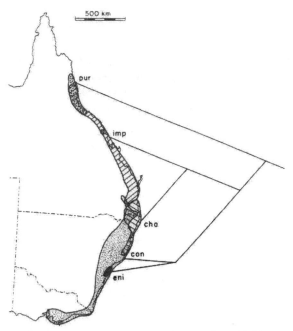

FIGURE 1. Phylogeny and Australian distribution of five species in the *Rhytidoponera impressa* group (*purpurea, impressa, chalybaea, confusa,* and *enigmatica*). (See also FIGURE 3.) *R. purpurea* also occurs in New Guinea; populations of *R. chalybaea* occurring farther south, in the vicinity of Sydney, and in New Zealand, are not included since they appear to represent recent introductions (Ward 1980a). An earlier phenetic analysis of electrophoretic data (Ward 1980b), where *populations* were the terminal units, gave different results from the species-level cladogram given here: it showed *R. enigmatica* and *R. impressa* diverging from populations of *R. chalybaea*. Reanalysis of those population data using cladistic methods gives results closer to the species-level cladogram, i.e. *R. enigmatica* is shown to be more closely related to *R. confusa* than to *R. chalybaea*, but *R. chalybaea* remains paraphyletic relative to *R. impressa* (Ward unpublished).

is a possible exception in that its range is enclosed within that of *R. confusa*.

Four of the five species in the *R. impressa* group have winged queens (in contrast to most members of this genus) so the partial overlap in ranges is not surprising. Nevertheless species with winged queens may exhibit limited dispersal. Moffett's (1985 1986) phylogenetic analyses of the species in the Indo-

Malayan genera *Myrmoteras* and *Acanthomyrmex*, in which the known queens are winged, revealed a pattern in which closely related pairs or clusters of species were almost invariably distributed allopatrically.

By contrast, closely related species in the New World genus *Pseudomyrmex* are often sympatric (Ward 1985a). All the species in this large, (150+ species) genus have alate queens. The *P. pallidus* group, whose Nearctic representatives I recently revised taxonomically, includes closely related pairs (or triplets) of species in which one (or two) taxa are locally distributed and contained within the range of a more widespread species (Ward 1985a:241). *P. pallidus* and *P. ejectus* occur widely in southern United States and Central America while their siblings (*P. seminole, P. leptosus* for *P. pallidus;* and *P. brunneus* for *P. ejectus*) have much more restricted distributions. The relationships among these and allied taxa, as indicated by a cladistic analysis, are depicted in FIGURE 2. These results agree with my more subjective statements (Ward 1985a) except in the placement of *P. simplex.* The occurrence of sister species with such disparate geographical ranges suggests that, if speciation occurred allopatrically, it involved a small population of one member in the pair. This conforms to Mayr's (1954 1982) model of peripatric speciation, insofar as the divergent daughter population is geographically or ecologically marginal.

Consistent with this idea, the cladistic analysis shows the species of geographically-restricted distribution to be more morphologically divergent than their widespread relatives (FIGURE 2). In fact, there are no uniquely-derived features (autapomorphies) for the latter; they are "paraspecies" or "metaspecies" in cladistic terminology (Ackery and Vane-Wright 1985, Donoghue 1985) from which the derived daughter species can be said to have arisen. This may be a common situation in groups of ants in which speciation follows a peripatric pattern. It is difficult to accept the suggestion of some cladists (e.g. Donoghue 1985) that such metaspecies, at least when they are shown to be "positively paraphyletic", do not qualify as real biological entities. There is, for example, no reasonable alternative to treating *P. pallidus* as a good species; it cannot be subdivided into diagnosable units nor can it be justifiably combined with *P. seminole* and *P. leptosus* from which it is reproductively isolated. (I agree, however, that it would be useful to adopt a convention of flagging such metaspecies.) *P. seminole* and *P. leptosus* display socially parasitic traits which might favor an

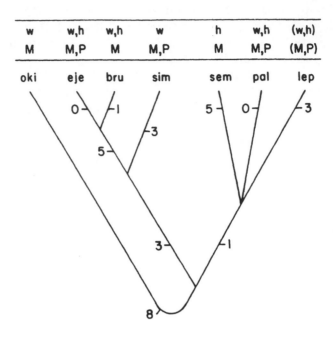

| Nest Sites | w | w,h | w,h | w | | h | w,h | (w,h) |
| No. of Queens | M | M,P | M | M,P | | M | M,P | (M,P) |

FIGURE 2. Cladogram depicting relationships among the Nearctic members of the *Pseudomyrmex pallidus* group (*ejectus, brunneus, simplex, seminole, pallidus, leptosus*). A Central American species, *P. oki*, which is closely related to, but not a member of, the *P. pallidus* group (Ward unpublished) was used as an outgroup to root the tree. Numbers indicate branch lengths; full data matrix is given in the Appendix. This tree has a consistency index of 0.862 (0.789 excluding autapomorphies of the original data matrix). Information is also given on nest site usage and numbers of queens within nests. w = nesting predominantly in woody trees and shrubs; h = nesting predominantly in herbaceous plants; w, h = commonly occupying both nest types; M = nests monogynous (single-queened); M,P = nests monogynous and polygynous.

hypothesis of nonallopatric speciation (Ward 1985a). I discuss the origin of social parasites in more detail later.

Genetic and Morphological Differences between Ant Species

The morphological and electrophoretic data on ants provide no indication of a qualitative shift in the kinds of genetic variation in the gradient from conspecific populations to closely related species. This is particularly demonstrated by the indeterminate status of allopatric populations of wingless-female *Rhytidoponera* (Ward 1984), which suggests a continuum of population divergence leading insensibly to speciation. In addition, my experience with both *Rhytidoponera* and the species-rich pseudomyrmecine genera, *Pseudomyrmex* and *Tetraponera*, indicates that the morphological characters which vary diagnostically between some species often exhibit the same range of variation *within* other species.

Another pattern, perhaps attributable to character displacement (Brown and Wilson 1956, Grant 1972), occurs repeatedly in *Pseudomyrmex.* Morphological characters which are diagnostic when a species occurs in association with a particular assemblage of congeners tend to vary outside those diagnostic bounds in other parts of its range (Ward 1985a). Similar cases have been reported in other ant genera, e.g. *Lasius* (Wilson 1955) and *Amblyopone* (Ward 1988).

There are some exceptions to these fickle patterns of variation. Some features of worker-ant morphology prove to be globally diagnostic for species, but their scarcity relative to the variable characters accounts, I think, for the continuing wealth of challenges in the α-taxonomy of ants.

Male genitalia provide reliable species-diagnostic characters in certain groups of ants, including *Pseudomyrmex* (Ward 1985 and unpublished), Ecitoninae (Borgmeier 1955), *Paratrechina* (Trager 1984), and *Myrmecia* (Browning unpublished). In some cases the genital features provide the most striking differences between otherwise very similar species. In these groups of ants the male genitalia may be prone to undergo rapid divergence during speciation, possibly as a consequence of sexual selection (Eberhard 1985). In general, however, the male genitalia of ants are rather conservative, i.e., relatively uniform among related species.

Karyotypic Variation in Ants

Karyotypic variation is dramatic within the family Formicidae, with haploid chromosome numbers ranging from n=1 (Crosland and Crozier 1986) to n~46 (Bishop and Crozier, in Taylor 1978). Most karyotype differences among related species of ants are explicable as Robertsonian rearrangements, pericentric inversions, and saltatory changes in constituitive heterochromatin (Imai *et al.* 1977). These rearrangements also appear as chromosome polymorphisms within a few species (Crozier 1970, Imai and Kubota 1975, Imai *et al.* 1977).

Karyotype differences among ant sibling species suggest an association between chromosomal change and speciation, but this suggestion must be tempered by the findings of intraspecific variation. As with differences in male genitalia, it is difficult to know whether chromosomal rearrangements drive divergence, or are merely the consequences of other changes (a general criticism of models of speciation invoking chromosomal rearrangements (Mayr 1982)).

There is no obvious correlation between karyotypic evolution and morphological divergence in ants (Imai *et al.* 1977). Some of the most extensive intra- and interspecific variation in chromosome number occurs within the subfamily Myrmeciinae (Imai *et al.* 1977, Crosland and Crozier 1986, Browning unpublished), which consists of a single relatively homogeneous genus *Myrmecia* (considered primitive in worker morphology relative to other ants). In contrast, genera in the phylogenetically advanced and morphologically diverse subfamily, Myrmicinae, tend to be conservative in chromosome number (Crozier 1975, Imai *et al.* 1977, Hauschteck-Jungen and Jungen 1983).

Conclusions: Ant Speciation

This evidence suggests that (1) allopatric differentiation is the predominant precursor to the establishment of reproductive isolation between ant populations (the extent to which reproductive isolation may be reinforced in sympatry is unclear), (2) such differentiation can occur on both a broad geographical scale and on a microgeographical level, and (3) there is no qualitative difference between genetic and morphological variation observed among closely related ant species and that seen within polytypic species.

These assertions are supported by the observations of (1) extensive intraspecific variation, often spanning the range of variation observed interspecifically among other, related species, (2) moderate to high levels of genetic differentiation among conspecific populations (high F_{ST}) coupled with low inbreeding coefficients ($F_{IS} \sim 0$), in those (admittedly few) species which have been examined electrophoretically, (3) the frequent allopatric distributions of closely-related taxa, especially among wingless-female ants, and (4) the disparate geographical ranges between some pairs of sister species.

SPECIATION AND SOCIAL BEHAVIOR IN ANTS

I address two questions here. First, are certain changes in social behavior closely associated with speciation in ants? This can be assessed indirectly by examining the intra- and interspecific distribution of social traits among taxa of known phylogenetic relationship. A paucity of intraspecific variation and frequent occurrences of sister species exhibiting contrasting modes of social organization would support the thesis that social changes are linked to speciation events. Another test would involve comparing the diversity of social behavior in speciose and species-poor taxa of the same age, but we have insufficient phylogenetic and behavioral information on ants to apply this second test at higher taxonomic levels, where it would be most powerful.

Second, to the extent that there is a tendency for changes in social behavior to be associated with speciation, are such changes likely to be partially responsible for the development of reproductive isolation? Here, I scrutinize the plausibility of arguments as well as the empirical support of comparative data.

What differences in colony organization are apparent within and among ant species? This question can be most readily answered for more conspicuous social traits such as the number of queens in a nest or the degree of physical-caste polymorphism. There is a dearth of comparative information about some of the more genetically relevant, but less conspicuous, traits such as frequency of polyandry, provenance of males, and relative reproductive contribution of queens (in nests with multiple queens). Moreover, reliable phylogenetic information is required to assess the extent to which changes in these variables are associated with speciation. Bearing these limitations in mind, I concentrate on the origin of two social traits in relation to ant speciation: polygyny and social parasitism.

Polygyny and Sibling Species of Ants

Most cases of polygyny in ants are attributable to the secondary addition of queens to a nest rather than colony foundation by more than one female (Hölldobler and Wilson 1977). In cases of extreme polygyny--unicolonial populations in which nests form a cooperating network over large areas and in

133

which dispersal occurs by fissioning--monogynous bottlenecks may seldom be achieved. Hölldobler and Wilson (1977) argued persuasively that certain ecological conditions, in particular ephemeral nest-sites and patchy habitats, favor the evolution of secondary polygyny. Crozier (1977a) suggested that when polygynous populations of ants arise they may in certain instances become reproductively isolated from their single-queened parent population. If this is true, and if polygyny arises fairly frequently, then monogynous-polygynous pairs of closely related species should be found. Hölldobler and Wilson (1977) claimed that such a pattern exists. I suspect, however, that at least some of these cases will prove to be artifacts of poor sampling or inadequate taxonomic knowledge and that *intraspecific* variation in this trait is more widespread than previously realized.

One example of a monogynous-polygynous species pair cited by Hölldobler and Wilson (1977) was "two sibling species of *P[seudomyrmex] flavidula* in Florida". *Pseudomyrmex flavidulus* belongs to the taxonomically difficult *P. pallidus* group, whose Nearctic representatives I recently revised (Ward 1985a). Floridian "*P. flavidulus*" comprise three phenotypically similar species, *P. seminole*, *P. simplex* and *P. pallidus*, of which the first is monogynous, the second is usually monogynous but rarely has polygynous colonies, and the third varies in the number of functional queens with some colonies being monogynous while others are polygynous with up to 22 mated queens in one nest (Ward 1985a). FIGURE 2 shows the inferred phylogenetic relationships among the Nearctic species of the *P. pallidus* group, together with information on nest sites and number of queens. Two points are worth noting: the *flavidulus*-like species do not form a monophyletic group, and the number of queens varies considerably within species in these and other members of the *P. pallidus* group (as, incidentally, do nesting habits). The picture of one or more monogynous-polygynous species pairs is certainly oversimplifed. There *are* predominantly polygynous species within the *P. pallidus* group; colonies of the South American species *P. ethicus* are usually multiple-queened. Elsewhere in the genus *Pseudomyrmex* some acacia-ants are highly polygynous (Janzen 1973).

Other supposed cases of monogynous-polygynous species pairs include "macrogyne" and "microgyne" forms of various *Myrmica* species (Brian 1983, Hölldobler and Wilson 1977, Pearson and Child 1980) but the status of these forms as good biological species remains unverified. They are not recognized as distinct species by Bolton and Collingwood (1975). There are reports of recently derived polygynous populations of the introduced fire ant, *Solenopsis invicta*, in the United States, but the detailed biochemical studies of Ross *et al.* (1987b) provide no evidence that these are reproductively isolated or genetically differentiated.

In the *Rhytidoponera impressa* group, three of the five species exhibit extensive variation in colony structure; polygynous, worker-reproductive colonies and monogynous, queenright colonies occur in the same species and frequently together in the same populations (Ward 1983). A fourth species, *R. purpurea*, is monogynous, while the fifth species, *R. enigmatica* is polygynous, but these two are not sister species (FIGURE 3). Rather, *R. enigmatica* is most closely related to *R. confusa* which has a variable gyne number. The polymorphic condition may be the ancestral one for the entire group (Ward 1983:296). Polygynous *Rhytidoponera* colonies are different from those of most ants in that the gynes are mated workers (gamergates) rather than deciduously winged queens.

Intraspecific variation in gyne number (from one to many queens) has been reported in other groups of ants (Hölldobler and Wilson 1977 and included references, Rosengren and Pamilo 1983, Pamilo and Rosengren 1984). Thus the number of functional egg-layers in a nest is most usefully viewed as a continuous variable, varying intraspecifically as well as interspecifically, and presumably reflecting (in many instances) variable selection pressures on gyne dispersal (Ward 1983, Rosengren and Pamilo 1983) and on worker tolerance of additional queens (Nonacs 1988); and perhaps complicated by conflict between individual- and colony-level selection. Detailed phylogenetic studies do not confirm the pattern of contrasting monogynous-polygynous pairs of sister species, and there seems to be ample scope for intraspecific evolution in this character. Insofar as ant speciation involves a small founding population (as discussed above) it is plausible that ecological pressures in the isolated habitat would favor increased levels of polygyny; and the process of differentiation could be accelerated by the local mating and reduced dispersal of queens. Whether this is of frequent importance in ant speciation is unclear, and it is not strongly supported by the evidence available to date.

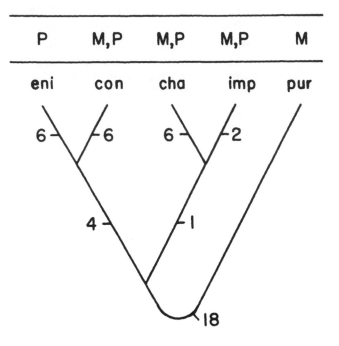

P	M,P	M,P	M,P	M
eni	con	cha	imp	pur

FIGURE 3. Cladogram of the five species in the *Rhytidoponera impressa* group (see also FIGURE 1), including lengths of the tree branches. Full data matrix is given in the Appendix. *R. purpurea*, the most morphologically and electrophoretically distinct member of the group (Ward 1980a 1980b) was used as an outgroup to root the tree. Consistency index of tree: 0.907. M = colonies monogynous and queenright; P = colonies polygynous and worker-reproductive; M,P = both colony types occur.

Social Parasitism in Ants

Interspecific social parasitism, in which one species utilizes the worker labor force of a different species, has evolved repeatedly in ants (Wilson 1971, Buschinger 1986). Dulotic or slave-making ants are a well-known example; these species obtain "slaves" by raiding other ant colonies for worker brood (Wilson 1975, Alloway 1980). A related type of social parasitism is exemplified by ants whose colony-founding queens invade non-conspecific nests, kill the host queen, and utilize the host workers on a temporary basis before replacing them with their own offspring (temporary social parasitism). In the more extreme case of workerless inquilinism, an invading parasitic queen usually tolerates the host queen (and her continued production of workers) while confining her own output to sexual reproductives. Several evolutionary pathways have been proposed for the development of social parasitism in ants (Buschinger 1970 1986, Wilson 1971, Alloway 1980, West-Eberhard 1987).

Some of these scenarios, inspired by comparative observations on intra- and interspecific variation in behavior, involve a graded series of steps leading to advanced dulosis and/or workerless inquilinism (e.g. Wilson 1971, Alloway 1980) and are not inextricably tied to one or more speciation events. But other authors, intrigued by the observation that socially parasitic ants often bear a closer taxonomic resemblance to their host species than to any other free-living species (a generalization known as "Emery's Rule"), suggest that such parasites originate from their hosts by sympatric speciation (Buschinger 1970 1986, Elmes 1978, Brian 1983, West-Eberhard 1987). This has been most forcefully argued for workerless inquilines; West-Eberhard (1987) proposes a model in which strong social and sexual selection act on small queens and males in an ancestral polygynous species, leading to assortative mating and a reproductively isolated workerless lineage.

What is the degree of generality of "Emery's Rule"? If the rule is stated strictly -- that social parasites are more closely related to their host species than to any other free-living taxa -- then it must be admitted that this *impression* of many investigators is based largely on phenetic resemblance rather than careful phylogenetic analysis (Emery 1909, Wheeler 1919, Wilson 1971, Buschinger 1981). Despite some recent progress (e.g. Francoeur 1981, Francoeur *et al.* 1985, Bolton 1988) considerable taxonomic confusion still prevails in groups of ants containing social parasites, such as the genus *Myrmica* and the tribe Leptothoracini. So most cases of Emery's Rule remain to be confirmed, although the looser generalization that social parasites exhibit close taxonomic affinities to their hosts has strong empirical support.

The preceding phylogenetic analysis of the Nearctic members of the *Pseudomyrmex pallidus* group reveals a case in which Emery's Rule does not hold. *P. leptosus* is a workerless social parasite (Ward 1985, Klein 1986) most likely derived from the metaspecies, *P. pallidus*, which attacks a less closely related species, *P. ejectus* (FIGURE 2). West-Eberhard (1987) mentions other cases in which Emery's Rule appears to break down, although the evidence is based on phenetic similarity (when a social parasite shows greater overall resemblance to a non-host species than to its host). West-Eberhard (1987) interprets such cases as possible instances of secondary host transfer, following sympatric speciation.

Even if Emery's Rule is tentatively accepted, there is no reason why this should *require* a sympatric speciation event. Wilson (1971) gives a plausible alternative involving genetic divergence of two geographically isolated populations and development of parasitic habits upon reinvasion of one another's ranges. Pearson (1981) suggests that competition between such siblings upon secondary contact could accentuate the divergence and favor a microgynous parasite as one member of the pair. Of course, such an allopatric model is readily extended to cases in which Emery's Rule does not strictly apply.

West-Eberhard (1987) contends that for the pattern described by Emery's Rule to be generated by allopatric speciation, the intermediate, allopatric transition populations would have to be either extinct or unusually divergent. Otherwise they, rather than the host species, would bear the closest taxonomic resemblance to the parasite. However, this overlooks the fact that in the allopatric model the geographical isolate is phyletically continuous with the social parasite that develops upon secondary sympatry, assuming such secondary contact involves complete overlap of populations. This is a reasonable expectation if speciation occurs in a peripatric fashion.

Studies of the genus *Myrmica* do provide suggestive evidence for a sympatric origin of divergence in the evolution of social parasitism (Brian and Brian 1955, Elmes 1978, Pearson and Child 1980). In several species of *Myrmica* small queens (microgynes) are associated with normal queens (macrogynes) and their status varies from apparently conspecific forms to clearly differentiated workerless parasites. Some of the latter are placed in the same genus (e.g. Elmes 1978, Francoeur 1981), while highly modified parasitic taxa have been assigned to their own genus, *Sifolinia*. *Sifolonia* was recently synonymized with *Myrmica* (Bolton 1988), obviously a desirable move as *Myrmica* would otherwise be paraphyletic. Most problematic are the closely similar forms, e.g. microgynes and macrogynes of *Myrmica ruginodis* (Brian and Brian 1955) and *M. rubra* (Pearson 1981), for which there is some indication of ecological and genetic differentiation. These ants deserve increased taxonomic and genetic scrutiny, especially with regard to the critical issue of assortative mating between the two morphs.

CONCLUSIONS

A theme throughout this paper is that problems of social evolution which involve comparative studies among taxa are critically dependent on detailed systematic knowledge. There are two aspects to this: species recognition and the assessment of phylogenetic relationships among these species. For outbreeding sexual organisms the biological species concept is clearly the most realistic model, yet considerable work may be required to delineate such species in ants and it may be difficult to apply the concept over a broad geographical area. Cladistic (minimum-length tree estimation) methods can be used for phylogenetic inference, but it should be borne in mind that the resulting cladograms contain groupings of varying empirical support. The case histories presented here, which involve social organization in *Pseudomyrmex* and *Rhytidoponera*, are based upon the most parsimonious interpretations of the available character sets (see Appendix), but the cladograms should be viewed critically, especially where tree branch lengths are short, or if alternative topologies of almost equal economy are possible. Even more caution is in order when phylogenetic analyses have not been attempted; in anticipation of future studies on social parasites it should be noted that morphological convergence associated with parasitism may partially obscure the evolutionary history of the taxa concerned (West-Eberhard 1987). For these studies it may be desirable to concentrate on alternative character systems such as male genitalia, allozymes, or karyotypes.

Some general conclusions that emerge from this assessment of the connection between social change and speciation in ants are: (1) Speciation in ants typically proceeds by allopatric differentiation of populations. This conclusion follows from observations on patterns of geographical variation and on the distributions of closely related taxa, especially among wingless-female ants. (2) Differentiation may proceed on a large geographical scale or it may involve a small, local population (peripatric speciation). Phylogenetic studies of *Rhytidoponera* and *Pseudomyrmex* provide evidence of both kinds of events. (3) The evolution and modification of social traits, such as the number of functional queens in a colony, is not closely linked to the origin of reproductive isolation between taxa. The diversity of social behavior exhibited by ants is compatible with phyletic change within lineages, and there is no indication that evolution is largely confined to speciation events. These assertions are supported by observations of intraspecific variation in social organization of ant colonies (particularly with respect to levels of polygyny), by the frequent persistence of such polymorphism across related taxa, and by the scarcity (up to this point) of phylogenetically well-documented cases of sister species with strongly contrasting social behavior. (4) The evolution of socially parasitic ants, including those that bear a close taxonomic resemblance to their host, does not *require* a sympatric mode of speciation. Moreover, there are difficulties inherent in establishing and maintaining assortative mating among the free-living and proto-parasitic morphs in a single population. The strongest empirical indication that something like this can occur comes from studies of microgyne and macrogyne forms in certain *Myrmica* species.

Acknowledgements

This work was supported by the University of California and by NSF BSR-8507865. I thank Bill Brown, Alfred Buschinger, Jim Carpenter, Ross Crozier, Jack Longino, Steve Shattuck and Mary Jane West-Eberhard for helpful discussions or comments on the manuscript.

LITERATURE CITED

Ackery, P. R. and R. I. Vane-Wright. 1984. *Milkweed butterflies: their cladistics and biology.* Cornell Univ. Press:Ithaca, New York.

Alloway, T. M. 1980. The origins of slavery in leptothoracine ants (Hymenoptera: Formicidae). *Am. Nat.* 115:247-261.

Avery, P. J. 1984. The population genetics of haplo-diploids and X-linked genes. *Genet. Res.* 44:321-342.

Bolton, B. 1980. The ant tribe Tetramoriini (Hymenoptera: Formicidae). The genus *Tetramorium* Mayr in the Ethiopian zoogeographical region. *Bull. Br. Mus. Nat. Hist. (Entomol.)* 40:193-384.

Bolton, B. 1986. Apterous females and shift of dispersal strategy in the *Monomorium salomonis*-group (Hymenoptera:Formicidae). *J. Nat. Hist.* 20:267-272.

Bolton, B. 1988. A new socially parasitic *Myrmica*, with a reassessment of the genus (Hymenoptera: Formicidae). *Syst. Entomol.* 13:1-11.

Bolton, B. and C. A. Collingwood. 1975. *Hymenoptera, Formicidae. Handbooks for the identification of British insects.* Vol. VI. Part 3(c).

Borgmeier, T. 1955. Die Wanderameisen der neotropischen Region. *Studia Entomol. (n.s.)* 3:1-720.

Brian, M. V. and A. D. Brian. 1955. On the two forms macrogyna and microgyna of the ant *Myrmica rubra* L. *Evolution* 9:280-290.

Brian, M. V. 1983. *Social insects, ecology and behavioural biology.* Chapman and Hall:London.

Brown, W. L. 1957. Centrifugal speciation. *Quart. Rev. Biol.* 32:247-277.

Brown, W. L. 1958. Contributions towards a reclassification of the Formicidae. II. Tribe Ectatommini. *Bull. Mus. Comp. Zool.* 118:175-362.

Brown, W. L. 1975. Contributions toward a reclassification of the Formicidae. V. Ponerinae, tribes Platythyreini, Cerapachyini, Cylindromyrmecini, Acanthostichini, and Aenictogitini. *Search Agric.* 5:1-116.

Brown, W. L. and E. O. Wilson. 1956. Character displacement. *Syst. Zool.* 5:49-64.

Brothers, D. J. 1975. Phylogeny and classification of the aculeate Hymenoptera, with special reference to Mutillidae. *Univ. Kansas Sci. Bull.* 50:483-848.

Buschinger, A. 1970. Neue Vorstellungen zur Evolution des Sozialparasitismus und der Dulosis bei Ameisen (Hym. Formicidae). *Biol. Zentralbl.* 88:273-299.

Buschinger, A. 1981. Biological and systematic relationships of social parasitic Leptothoracini from Europe and North American. In *Biosystematics of social insects*, P. E. Howse and J.-L. Clément, eds. Systematics Association Special Volume No. 19. Academic Press:London pp. 211-222.

Buschinger, A. 1986. Evolution of social parasitism in ants. *Trends Ecol. Evol.* 1:155-160.

Charlesworth, B., J. A. Coyne and N. H. Barton. 1987. The relative rates of evolution of sex chromosomes and autosomes. *Am. Nat.* 130:113-146.

Charlesworth, B., R. Lande and M. Slatkin. 1982. A neo-Darwinian commentary on macroevolution. *Evolution* 36:474-498.

Creighton, W. S. 1950. The ants of North America. *Bull. Mus. Comp. Zool.* 104:1-585.

Crosland, M. W. J. and R. H. Crozier. 1986. *Myrmecia pilosula*, an ant with only one pair of chromosomes. *Science* 231:1278.

Crozier, R. H. 1970. Pericentric rearrangement polymorphism in a North American dolichoderine ant (Hymenoptera: Formicidae). *Can. J. Genet. Cytol.* 12:541-546.

Crozier, R. H. 1975. *Animal cytogenetics 3. Insecta 7. Hymenoptera.* Gebrüder Borntraeger:Berlin.

Crozier, R. H. 1977a. Evolutionary genetics of the Hymenoptera. *Ann. Rev. Entomol.* 22:263-288.

Crozier, R. H. 1977b. Genetic differentiation between populations of the ant *Aphaenogaster "rudis"* in the southeastern United States. *Genetica* 47:17-36.

Crozier, R. H. 1979. Genetics of sociality. In *Social insects, Vol. I,* Hermann, H. R., ed.. Academic Press:New York pp. 223-286.

Crozier, R. H. 1981. Genetic aspects of ant evolution. In *Evolution and speciation. Essays in honor of M. J. D. White,* W. R. Atchley and D. Woodruff, eds. Cambridge Univ. Press:Cambridge pp. 356-370.

Crozier, R. H., P. Pamilo, R. W. Taylor and Y. C. Crozier. 1986. Evolutionary patterns in some putative Australian species in the ant genus *Rhytidoponera. Aust. J. Zool.* 34:535-560.

Donoghue, M. J. 1985. A critique of the biological species concept and recommendations for a phylogenetic alternative. *Bryologist* 88:172-181.

Eberhard, W. G. 1985. *Sexual selection and animal genitalia.* Harvard Univ. Press:Cambridge, Mass.

Elmes, G. W. 1978. A morphometric comparison of three closely related species of *Myrmica* (Formicidae), including a new species from England. *Syst. Entomol.* 3:131-145.

Emery, C. 1909. Über den Ursprung der dulotischen, parasitischen und myrmekophilen Ameisen. *Biol. Centralbl.* 29:352-362.

Emery, C. 1925. Hymenoptera. Fam. Formicidae. Subfam. Formicinae. In *Genera insectorum,* P. Wytsman, ed. Fasc 183. 320 p.

Francoeur, A. 1973. Révision taxonomique des espèces néarctiques du groupe *fusca,* genre *Formica* (Formicidae, Hymenoptera). *Mém. Soc. Entomol. Québec* 3:1-316.

Francoeur, A. 1981. Le groupe Néarctique *Myrmica lampra* (Formicidae, Hymenoptera). *Can. Ent.* 113:755-759.

Francoeur, A., R. Loiselle and A. Buschinger. 1985. Biosystématique de la tribu Leptothoracini (Formicidae, Hymenoptera). 1. Le genre *Formicoxenus* dans la région Holarctique. *Naturaliste Can. (Rev. Écol. Syst.)* 112:343-403.

Futuyma, D. 1987. On the role of species in anagenesis. *Am. Nat.* 130:465-473.

Gotwald, W. H. 1982. Army ants. In *Social insects, Vol. IV*, H. R. Hermann, ed. Academic Press:New York pp. 157-254.

Gould, S. J. and N. Eldredge. 1977. Punctuated equilibria: the tempo and mode of evolution reconsidered. *Paleobiology* 3:115-151.

Grant, P. 1972. Convergent and divergent character displacement. *Biol. J. Linn. Soc.* 4:39-68.

Halliday, R. B. 1981. Heterozygosity and genetic distance in sibling species of meat ants (*Iridomyrmex purpureus* group). *Evolution* 35:234-242.

Hartl, D. L. 1972. A fundamental theorem of natural selection for sex linkage or arrhenotoky. *Am. Nat.* 106:516-524.

Hauschteck-Jungen, E. and H. Jungen. 1983. Ant chromosomes. II. Karyotypes of western palearctic species. *Insectes Sociaux* 30:149-164.

Hölldobler, B. and E. O. Wilson. 1977. The number of queens: an important trait in ant evolution. *Naturwissenschaften* 64:8-15.

Imai, H. T. and M. Kubota. 1975. Chromosome polymorphism in the ant *Pheidole nodus. Chromosoma* 51:391-399.

Imai, H. T. and R. W. Taylor. 1986. The exceptionally low chromosome number n = 2 in an Australian bulldog ant, *Myrmecia piliventris* Smith (Hymenoptera: Formicidae). *Ann. Report Natl. Inst. Genet.* 36:59-61.

Imai, H T., R. H. Crozier and R. W. Taylor. 1977. Karyotype evolution in Australian ants. *Chromosoma* 59:341-393.

Janzen, D. H. 1973. Evolution of polygynous obligate acacia-ants in western Mexico. *J. Anim. Ecol.* 42:727-750.

Klein, R. W. 1986. A workerless inquiline in *Pseudomyrmex. Abstracts Tenth Congr. Int. Union Study Soc. Insects, Munich*, p. 206.

Mayr. E. 1954. Change of environment and evolution. In *Evolution as a process*, J. Huxley, A. C. Hardy and E. B. Ford, eds. Allen and Unwin:London pp. 157-180.

Mayr. E. 1982. Processes of speciation in animals. In *Mechanisms of speciation*, C. Barigozzi, Alan Liss, Inc.:New York pp. 1-19.

Moffett, M. W. 1985. Revision of the genus *Myrmoteras* (Hymenoptera: Formicidae). *Bull. Mus. Comp. Zool.* 151:1-53.

Moffett, M. W. 1986. Revision of the myrmicine genus *Acanthomyrmex* (Hymenoptera: Formicidae). *Bull. Mus. Comp. Zool.* 151:55-89.

Nonacs, P. 1988. Queen number in colonies of social Hymenoptera as a kin-selected adaptation. *Evolution* 42:566-580.

Oster, G. F. and E. O. Wilson. 1978. *Caste and ecology in the social insects.* Princeton Univ. Press:Princeton, NJ.

Pamilo, P. 1983. Genetic differentiation within subdivided populations of *Formica* ants. *Evolution* 37:1010-1022.

Pamilo, P. and R. Rosengren. 1984. Evolution of nesting strategies in ants: genetic evidence from different population types of *Formica* ants. *Biol. J. Linn. Soc.* 21:331-348.

Pearson, B. and A. R. Child. 1980. The distribution of an esterase polymorphism in macrogynes and microgynes of *Myrmica rubra* Latreille. *Evolution* 34:105-109.

Pearson, B. 1981. The electrophoretic determination of *Myrmica rubra* microgynes as a social parasite: possible significance in the evolution of ant social parasites. In *Biosystematics of social insects*, P. E. Howse and J-.L. Clément, eds. Systematics Association Special Volume No. 19. Academic Press:London pp. 75-84.

Rosengren, R. and P. Pamilo. 1983. The evolution of polygyny and polydomy in mound-building *Formica* ants. *Acta Entomol. Fennica* 42:65-77.

Ross, K. G. and D. J. C. Fletcher. 1985. Comparative study of genetic and social structure in two forms of the fire ant, *Solenopsis invicta* (Hymenoptera: Formicidae). *Behav. Ecol. Sociobiol.* 17:349-356.

Ross, K. G., R. K. Van der Meer, D. J. C. Fletcher and E. L. Vargo. 1987a. Biochemical phenotypic and genetic studies of two introduced fire ants and their hybrid (Hymenoptera: Formicidae). *Evolution* 41:280-293.

Ross, K. G., E. L. Vargo and D. J. C. Fletcher. 1987b. Comparative biochemical genetics of three fire ant species in North America, with special reference to the two social forms of *Solenopsis invicta* (Hymenoptera: Formicidae). *Evolution* 41:979-990.

Taylor, R. W. 1978. *Nothomyrmecia macrops*: a living-fossil ant rediscovered. *Science* 201:979-985.

Trager, J. C. 1984. A revision of the genus *Paratrechina* (Hymenoptera: Formicidae) of the continental United States. *Sociobiology* 9:51-162.

Vepsäläinen, K. and B. Pisarski. 1981. The taxonomy of the *Formica rufa* group: chaos before order. In *Biosystematics of social insects*, P. E. Howse and J-.L. Clément, eds. Systematics Association Special Volume No. 19. Academic Press:London pp. 27-36.

Ward, P. S. 1980a. A systematic revision of the *Rhytidoponera impressa* group(Hymenoptera: Formicidae) in Australia and New Guinea. *Aust. J. Zool.* 28:475-498.

Ward, P. S. 1980b. Genetic variation and population differentiation in the *Rhytidoponera impressa* group, a species complex of ponerine ants (Hymenoptera: Formicidae). *Evolution* 34:1060-1076.

Ward, P. S. 1983. Genetic relatedness and colony organization in a species complex of ponerine ants. I. Phenotypic and genotypic composition of colonies. *Behav. Ecol. Sociobiol.* 12:285-299.

Ward, P. S. 1984. A revision of the ant genus *Rhytidoponera* (Hymenoptera: Formicidae) in New Caledonia. *Aust. J. Zool.* 32:131-175.

Ward, P. S. 1985a. The Nearctic species of the genus *Pseudomyrmex* (Hymenoptera: Formicidae). *Quaest. Entomol.* 21:209-246.

Ward, P. S. 1985b. Taxonomic congruence and disparity in an insular ant fauna: *Rhytidoponera* in New Caledonia. *Syst. Zool.* 34:140-151.

Ward, P. S. 1988. Mesic elements in the western Nearctic ant fauna: taxonomic and biological notes on *Amblyopone, Proceratium,* and *Smithistruma. J. Kansas Entomol. Soc.* 61:102-124.

West-Eberhard, M. J. 1986. Alternative adaptations, speciation, and phylogeny. *Proc. Natl. Acad. Sci. USA.* 83:1388-1392.

West-Eberhard, M. J. 1987. Sympatric speciation in socially parasitic ants. Unpublished manuscript.

Wheeler, W. M. 1917. The phylogenetic development of subapterous and apterous castes in the Formicidae. *Proc. Natl. Acad. Sci. USA* 3:109-117.

Wheeler, W. M. 1919. The parasitic Aculeata, a study in evolution. *Proc. Amer. Phil. Soc.* 58:1-40.

Wheeler, W. M. 1923. The occurrence of winged females in the ant genus *Leptogenys* Roger, with descriptions of new species. *Amer. Mus. Novitates* 90:1-16.

Wheeler, W. M. 1934. A second revision of the ants of the genus *Leptomyrmex. Bull. Mus. Comp. Zool.* 77:69-118.

Wilson, E. O. 1955. A monographic revision of the ant genus *Lasius. Bull. Mus. Comp. Zool.* 113:1-201.

Wilson, E. O. 1958. Studies on the ant fauna of Melanesia, I. The tribe Leptogenyini, II. The tribes Amblyoponini and Platythyreini, III. *Rhytidoponera* in western Melanesia and the Moluccas, IV. The tribe Ponerini, V. The tribe Odontomachini. *Bull. Mus. Comp. Zool.* 118:101-153; 119:303-371; 120:483-510.

Wilson, E. O. 1971. *The insect societies.* Harvard Univ. Press:Cambridge, Mass.

Wilson, E. O. 1975. *Leptothorax duloticus* and the beginnings of slavery in ants. *Evolution* 29:108-119.

Wilson, E. O. 1987. The earliest known ants: an analysis of the Cretaceous species and an inference concerning their social organization. *Paleobiology* 13:44-53.

Wing, M. W. 1968. Taxonomic revision of the Nearctic genus *Acanthomyops* (Hymenoptera: Formicidae). *Mem. Cornell Univ. Agric. Exp. Station* 405:1-173.

APPENDIX

Here I provide the character state data from which cladograms (FIGURES 1-3) were derived. The cladograms represent the most parsimonious trees obtained with the ALLTREES option on David Swofford's PAUP program. For both character sets "?" signifies inapplicable or missing data. The first species listed was used as an outgroup to root the tree.

(1) *Rhytidoponera impressa* group

purpurea	001100010000010000101010001000000000000
confusa	111001100101010011010001010211100001010
engimatica	010010000101010110010001110211110021110
chalybaea	010010001010111011010101010111001010010
impressa	010011000000001001001000101011100011?010

Characters *1-27* (first 27 columns) refer to the presence (1) or absence (0) of electrophoretically detectable alleles at the following seven loci (see also Ward, 1980b): EST-3 (1-4), EST-4 (5-8), AMY (9-15), A0-2 (16-19), MDH (20-22), SOD-1 (23-25), PGM (26-27). An allele was considered present if its weighted gene frequency over all conspecific populations sampled exceeded 0.05. The remaining characters are morphological and life history traits, as follows. *28* Worker scapes (0) not widened basally, (1) slightly widened basally, (2) much expanded basally. *29* Worker lateral occipital lobes (0) broadly rounded, (1) narrowly rounded. *30* Worker subpetiolar process (0) with long spine-like tooth; not ventrally expanded posterior to the tooth, (1) with shorter tooth; ventrally expanded posterior to this. *31* Worker subpetiolar tooth (0) spine-like, (1) more blunt-toothed. *32* Worker, striation on fourth abdominal tergum (0) transverse, (1) longitudinal or transversely arched. *33* Worker, striation on fourth abdominal tergum (0) laterally obsolete (1) extending more or less to the lateral margin. *34* Worker color (0) iridescent blue, green, or purple (1) non-iridescent brown. *35* Worker iridesence (0) green-purple, with brassy gaster, (1) blue or blue-purple, (2) blue with grey-green reflections. *36* Queen caste (0) present, (1) absent. *37* Worker reproduction (0) absent, (1) present. *38* Reproductives released in (0) spring, (1) fall. Character 35 was considered unordered.

(2) *Pseudomyrmex pallidus* group (with *P. oki* as outgroup)

oki	00012000000100000000
ejectus	00002110111000112001
brunneus	00002111111000112001
simplex	00002000211210112010
seminole	01120100110210111100
pallidus	00011100110210111000
leptosus	100111?0210211111000

The characters and character states are as follows (following the columns from left to right). *1* Worker and queen palpal formula (0) 5,3 (1) 4,3. *2* Queen size (0) small (HL < 1.15), (1) large (HL > 1.25) *3* Worker and queen, median clypeal lobe (0) without median protuberance, (1) with slight (worker) or strong (queen) median protuberance. *4* Queen, frontal carinae (0) closely contiguous (MFC usually < 0.020), (1) moderately close (MFC 0.020-0.040), (2) well separated (MFC > 0.060). *5* Queen eye length (0) short (REL < 0.45), (1) medium (REL 0.45-0.50), (2) long (REL > 0.50). *6* Worker and male, forefemur (0) broadened (worker FI > 0.45, male FI > 0.36), (1) not broadened. *7* Worker, basal face of propodeum (0) subequal to or longer than, declivitous face, not preceded by a deep metanotal groove, (1) shorter than declivitous face and preceded by a deeply impressed metanotal groove. *8* Worker and queen, petiole and postpetiole (0) not notably widened, (1) much widened posterolaterally (PPWI > 1.25). *9* Worker and queen, frons sculpture (0) densely punctulate, opaque, (1) coriarious and more sparsely punctulate, sublucid, (2) predominantly smooth and shiny. *10* Worker and queen, standing pilosity (0) common on mesonotum and propodeum, (1) lacking (worker) or sparse (queen) on mesonotum and propodeum. *11* Worker and queen, pubesence on fourth abdominal tergum (0) moderately dense, (1) relatively sparse, not obscuring sheen of the integument. *12* Worker and queen, head (0) dark brown, (1) medium brown, (2) orange-brown. *13* Worker and queen, gaster (0) dark brown, (1) light orange-brown. *14* Worker caste (0) present, (1) absent. *15* Male, posterior margin of pygidium

147

(0) directed ventrally, (1) strongly recurved forward. *16* Male, inner face of paramere (0) with a broadly attached lamella, (1) without a broadly attached lamella (a basally attached free-standing protuberance is present). *17* Male paramere (0) little enlarged caudally, (1) with a rounded apicodorsal lobe, (2) with a large apicodorsal lobe preceded by a small spine. *18* Male paramere, caudal end (0) simple, (1) dorsally bicarinate, with a central lunule. *19* Male, inner caudal margin of paramere, seen in dorsal view (0) concave, (1) straight. *20* Male hypopygium (0) without a median ventral protuberance, (1) with a median ventral protuberance.

CHAPTER 8

REPRODUCTIVE AND SOCIAL STRUCTURE IN POLYGYNOUS FIRE ANT COLONIES

Kenneth G. Ross

Recently a considerable interest has grown in polygynous (multiple-queen) societies of highly eusocial insects (Hölldobler and Wilson 1977, Jeanne 1980, Brian 1983, Pamilo and Rosengren 1984, Thorne 1984, Fletcher and Ross 1985). This interest is primarily motivated by the realization that the origin and subsequent development of such societies from their presumed monogynous (single-queen) ancestors presents many of the same theoretical challenges as does the origin of eusociality. Evolutionary explanations of polygyny must resolve the paradox of why queens should risk some loss of their personal reproduction by associating in groups. Any insights gained concerning polygyny may thus also be of some relevance to understanding the inception of eusocial behavior in insects.

How reproductive effort is distributed among queens is an especially significant issue relating to the evolution of polygyny in highly eusocial forms. Inequities in reproductive success among associating queens alter the individual cost-benefit ratios that are presumed to be important in promoting polygyny, thus introducing a further element of complexity in attempts to interpret the roles of various factors such as kin selection or mutualism in the evolution of this social organization (Crozier 1979, Pamilo 1982, Ross and Fletcher 1985a, Nonacs 1988). Variation in reproductive success may be interpreted as evidence for social competition for reproductive privileges (e.g., West-Eberhard 1981, 1983, Fletcher and Ross 1985), and such reproductive competition may be a destabilizing influence on

Department of Entomology, University of Georgia, Athens, GA 30602, USA polygyny. Such reproductive competition is likely to be most prominent in polygynous groups in which queens are not close kin because each queen's personal reproduction is its major or sole fitness component.

Reproductive effort in polygynous ants is frequently assumed to be distributed rather evenly among queens (Hölldobler and Wilson 1983), particularly in highly polygynous forms where colony queen number is high. These queens have thus been viewed as comprising a relatively homogeneous class of reproductive females. However, more recent data based on observations of queen interactions and fecundity (e.g., Plateaux 1981, Evesham 1984, Hölldobler and Carlin 1985, Heinze and Buschinger 1987) indicate that nestmate queens may not have equivalent reproductive competency. It has been the objective of my colleagues, D. J. C. Fletcher and E. L. Vargo, and me to reach a better understanding of reproductive roles in polygynous fire ant colonies, with the eventual aim of determining what influence inequities in individual reproductive success may have in structuring these societies and directing their evolution. In this review I focus on some of our recently completed work.

Our research organism, *Solenopsis invicta*, occurs in both monogynous and polygynous forms in the southeastern United States, thus providing a valuable system for comparative study of these two social organizations. This ant is readily cultured and manipulated in the laboratory and there is a growing volume of background data on its population genetics. This genetic work has yielded of genetic markers useful for estimating relatedness, determining ploidy, and measuring individual reproductive success. A further virtue of this species as a model system is that workers of *S. invicta* are obligately sterile. Thus, consideration of reproductive roles is somewhat simplified by the limitation of reproduction to the queen.

THE COST OF POLYGYNY

A negative association exists between colony queen number and mean queen fecundity in a diversity of social Hymenoptera (e.g., Michener 1964, Wilson 1971, Yamauchi *et al.* 1982, Mercier *et al.* 1985). This so-called "reproductivity effect" has been studied in *S. invicta* in both the laboratory (Fletcher *et al.* 1980, Greenberg *et al.* 1985, Vargo and Fletcher 1988) and in the field (Vargo and Fletcher 1988). The field study demonstrated that the weight of dealate queens, which is highly correlated with individual fecundity (r=0.75 [n=115] and r=0.90 [n=30]; Vargo and Fletcher 1988, Ross 1988, respectively), decreased dramatically as queen number increased between one and 50. Significantly, queens from mature monogynous colonies of *S. invicta* generally weigh more than even the heaviest of queens from polygynous colonies (Vargo and Fletcher 1988). This suggests that, relative to successful monogynous conspecifics, queens participating in polygynous associations sacrifice some personal reproductive output, the magnitude of which increases as the number of queens in a colony increases.

THREE MAJOR REPRODUCTIVE ROLES CHARACTERIZE POLYGYNOUS ASSOCIATIONS OF *S. INVICTA* QUEENS

Given that queen association seems to entail some cost to average individual reproduction, we wanted to determine how reproductive effort is distributed among queens comprising polygynous groups. If some queens have higher reproductive success than do their nestmates, then the cost of polygyny cannot be equivalent across all individuals. We thus looked for evidence of disparity in reproductive success among nestmate queens.

In fact, even at a rather coarse level of analysis in which only mating status and type of offspring produced are considered, dealate queens comprise three distinct reproductive classes, among which differences in reproductive success are probably significant.

Uninseminated Queens

A rather striking feature of polygynous *S. invicta* colonies is the high proportion of dealate queens which are uninseminated (Fletcher *et al.* 1980, Fletcher 1983, Vargo and Fletcher 1987, 1988). About 30% of the queens found in polygynous populations are inseminated, although their representation in individual nests is highly variable (range=0 to 51%, n=107 nests; Vargo and Fletcher 1987, 1988). The origin of these queens remains a mystery largely because of our ignorance of the mating and dispersal biology of polygynous *S. invicta*. They may represent an ontogenetic stage (i.e., young queens that have not yet mated), they may result from dealation in the nest during periods of low levels of queen inhibitory pheromone (see Fletcher 1986, Fletcher and

151

Blum 1983a,b), or they may be queens unable to find mates due to the relative paucity of haploid males available in these populations (see below, also Vargo and Fletcher 1986, 1987). These possibilities are not mutually exclusive, nor are they exhaustive.

The conspicuous presence of these queens raises questions about their reproductive role in polygynous societies. Being uninseminated, their reproductive contribution can consist only of haploid males, and, given their large numbers, it may be reasonable to hypothesize that they produce a significant proportion of males in polygynous populations. However, data from four such populations indicate that only a small percentage of males are haploid ($\overline{x}$ =6.8%, range=0 to 90% per nest, n=55 nests; Ross and Fletcher 1985b), the remainder being diploid males produced by mated queens (see below). Nonetheless, as haploid males are the male component of the Mendelian population (that is, are the only functional male sources for genetic contribution to subsequent generations), we undertook to determine what role unmated dealate queens play in producing them.

Vargo and Fletcher (1988) found that unmated dealate queens in polygynous nests of S. invicta are invariably of lower average weight (and thus less fecund) than their mated nestmates, with significant differences in nearly three-fourths of the field colonies studied. Furthermore, the average proportion of viable eggs produced, as assessed by the ability of eggs to form embryos (Voss 1985), is much lower for unmated than for mated queens, both for queens held in isolation in the laboratory (TABLE 1) and for queens recently removed from colonies in the field (E. Vargo and K. Ross, unpublished). The natural history and physiological correlates of production of non-embryonated eggs in polygynous S. invicta are poorly known, but these eggs presumably are fed to brood and thus likely serve a trophic function in the colony (see Fletcher and Blum 1983b). In any case, it is clear that they are incapable of further development (Voss et al. 1988) and that their presence is not related to the direct production of offspring.

From these data it appears that unmated dealate queens are at a significant reproductive disadvantage when compared to their mated nestmates. Not only are they incapable of producing female sexual progeny, but they are relatively deficient in their fecundity and their capacity to produce viable, embryonated eggs. It seems rather unlikely that most unmated queens in polygynous

152

TABLE 1. Proportion of embryonated eggs produced by dealate polygynous queens of *S. invicta* held in isolation in the laboratory for several weeks. Differences between inseminated and uninseminated queens newly removed from field colonies are similar though less pronounced (Source: E. Vargo and K. Ross, unpublished).

Mating status	Proportion of eggs embryonated		N (queens)
	Mean ± SD	Range	
Inseminated	0.980 ± 0.013	0.856-1.0	53
Uninseminated	0.081 ± 0.068	0.0-0.535	60

populations contribute significantly to the Mendelian population.

Inseminated Queens

In addition to the conspicuous dichotomy of dealate queens based on mating status, the nature of their offspring produced divides mated queens into two distinct classes in polygynous *S. invicta*. These are mated queens that produce some proportion of diploid males and those that produce only diploid females from their fertilized eggs. Ross and Fletcher (1985b) argued that the occurrence of diploid-male-producing queens at significant frequencies in polygynous populations is related to a loss of genetic diversity associated with the early colonization of North America by *S. invicta*. The population bottleneck that likely accompanied founding of the North American population resulted in the loss of some alleles at the presumed single major sex locus and an attendant increase in the frequency of "matched" matings, that is, matings in which a queen mates with a male bearing an allele at the sex locus identical to one of hers (Adams *et al.* 1977). Queens of *S. invicta* and other fire ant species invariably mate with only one male (Ross and Fletcher 1985a, Ross *et al.* 1988), so that the percentage of diploid-male-producing queens in a population is identical to the percentage of matched matings. In Texas and Georgia populations this appears to be 15-20% of all matings (Ross and Fletcher 1985b, 1986, E. Vargo and K. Ross, unpublished). With a single locus system of sex determination, match-mated queens produce diploid male and diploid female zygotes in a 1:1 ratio (Crozier 1977).

The 15-20% of mated queens in polygynous populations that are diploid-male-producing queens appear to be at a major reproductive disadvantage compared to the remainder of mated queens that did not match-mate. This is because the contribution by diploid-male-producing queens to the pool of fertilized eggs in a colony is devalued by one-half, diploid males having no reproductive value (these are either sterile [Hung et al. 1974] or are likely to produce diploid or aneuploid sperm [Ross and Fletcher 1985b]). Thus only one-half of the zygotes produced by diploid-male-producing queens are potentially legitimate (female) sexual forms.

These results suggest that, based on mating status and the characteristic composition of offspring produced, there are three distinct classes of potential reproductive females in polygynous S. invicta colonies. They occur in approximately the following proportions in polygynous populations: uninseminated queens, 0.30; inseminated queens producing diploid males, 0.12; and inseminated queens producing only diploid females from fertilized eggs, 0.58. There are dramatic differences in the probability of achieving significant reproductive success among queens in these three classes, with inseminated queens that do not produce diploid males having an advantage over queens in the other two classes. The likely disparity in reproductive value between diploid-male-producing and other mated queens apparently arises as the result of a single chance event in a queen's lifetime, that is, the acquisition of a male mate with a particular allele at the major sex locus.

DIFFERENTIAL REPRODUCTION AMONG INDIVIDUAL QUEENS

Given the likely differences in reproductive output among queens of different reproductive classes, we have examined in greater detail the fine-scale social and reproductive structure of polygynous colonies by determining if different queens of the same type exhibit different reproductive potentials. Widely varying levels of fecundity and rates of egg embryonation among nestmate queens within each reproductive class had previously been found (Vargo and Fletcher 1988, Vargo and Ross unpublished data),1 suggesting the existence of such individual differences in reproductive output. Nestmate queens have low genetic relatedness to one another (Ross and Fletcher 1985a), a characteristic consistent with the emergence of significant competition for reproduction and differential reproductive success (Crozier 1979, West-Eberhard 1981, Fletcher and Ross 1985). Evidence for differential reproduction may suggest that social competition constitutes an important factor influencing social organization in polygynous fire ants.

Reproduction at the level of the individual queen was studied by determining the maternity of workers and female sexuals in multiple-queen laboratory colonies of known composition (Ross 1988). The four to six mother queens heading each study colony were chosen on the basis of the following criteria: (1) all originated from the same field colony, (2) all headed vigorous colonies in isolation prior to being re-united in the study colonies, and (3) each produced female offspring uniquely recognizable (relative to others produced in the colony) by virtue of their genotypes at two polymorphic enzyme loci. Each of the six colonies was studied over a period of five to 14 months. Samples of worker pupae and adult sexuals were taken regularly for electrophoretic determination of maternity.

Over short-term observation periods an often striking disparity in the apportionment of maternity of female sexuals existed. For instance, considering only the initial samples for each colony, over 84% of the total female sexual production (pooled over colonies) could be attributed to just eight of the 31 mother queens (26%). Significant inequity among mother queens in this respect characterized most of the samples from all six study colonies (Ross 1988). On the other hand, queen contributions to worker progeny were almost always more evenly distributed among mother queens in a colony. A measure of the variance in queen contribution, the average absolute deviation from the mean, was significantly higher for sexuals than for workers when these were compared on a sample-by-sample basis (Wilcoxon matched-pairs signed-ranks test, $p < 0.001$, n=18), confirming that mother queens generally contributed more equally to the pool of workers produced in a colony than to the pool of female sexuals. Furthermore, the degree of concordance between levels of worker and sexual maternity for individual queens was only moderate and variable (Kendall's tau=0.42, range -0.2 to 1.0, n=14), suggesting that queens which were more successful in having their daughters reared as sexuals during a given sample interval often did not produce a correspondingly large share of workers simultaneously.

Pronounced differences in long-term reproductive success among nestmate queens were also evident in this study, as is

155

TABLE 2. Apportionment of maternity of workers and female sexuals among mother queens over the course of the study of Ross (1988). Worker production is the mean of the values from all samples during which a queen was alive (thus these do not sum to one for each colony). Queens that died before completion of the study are indicated with asterisks. Their proportional production of sexuals for samples prior to their death is in parentheses.

Study colony	Mother queen	Proportion of worker offspring	Proportion of female sexual offspring
COL C1	C47	0.17	0.31
	C16	0.19	0.07
	C59	0.32	0.53
	C24*	0.39	0.04 (0.06)
	C15 [a]*	0.05	0.05 (0.07)
COL C2	C37	0.37	0.03
	C33*	0.31	0.23 (0.23)
	C54*	0.13	0.44 (0.48)
	C43[a]	0.22	0.30
COL D1	D37	0.09	0.02
	D36	0.06	0.01
	D25	0.06	0.02
	D33	0.43	0.51
	D53*	0.28	0.44 (0.69)
	D58	0.16	0
COL J1	J57	0.43	0.46
	J1*	0.16	0 (0)
	J35[a]	0.09	0
	J54*	0.07	0.49 (0.70)
	J42	0.37	0.05
COL R1	R43*	0.07	0.12 (0.15)
	R37*	0.12	0.56 (0.65)
	R16*	0.07	0.01 (0.01)
	R9	0.18	0
	R70	0.12	0.01
	R5	0.55	0.30
COL R2	R72*	0	0 (0)
	R34*	0.01	0 (0)
	R56[a]*	0.16	0.31 (0.32)
	R42	0.24	0.30
	R68	0.60	0.39

[a] Diploid-male-producing queens

documented in TABLE 2. Two factors contributed to this outcome: (1) several queens exhibited consistently low representation of their offspring among adult sexuals yet made significant contributions to the worker pool (e.g., Queens C24, C16, C37, D58, J42, R70, and R9 in TABLE 2); (2) several queens dominated production of female sexuals over short intervals and the production over these intervals comprised a large proportion of the entire production of sexuals over the course of the study.

A most intriguing finding of this study is that most of the mother queens that dominated production of reproductives in a given sample underwent a dramatic weight loss coincidentally, followed shortly by their death (TABLE 3). Particularly noteworthy are the cases of the diploid-male-producing queens C43 and R56, who dominated production of female reproductives during weight loss despite the fact that probably 50% of the fertilized eggs they laid were determined genetically to become diploid males (see Ross and Fletcher 1985b, Ross 1988).

The significance of the results of this latter study in relation to the evolution of reproductive roles and colony social organization in polygynous *S. invicta* is speculative pending similar investigations of natural colonies and elucidation of a mechanism of differential reproduction among queens in a single reproductive class. Ross (1988) suggests that over-representation of a queen's offspring in the pool of reproductive offspring does not arise solely from a relative fecundity advantage of successful queens, nor can it be attributed to a simple mechanism of kin recognition and discrimination by nurse workers during brood rearing. The weight loss syndrome manifested by most successful mother queens hints that such individuals might somehow bias their eggs toward development as sexuals, a phenomenon not without precedent in ants (see Wheeler 1986). If this view is correct, then the weight loss and death of such queens may result from increased investment in their eggs. More detailed studies of caste differentiation, brood-rearing behavior and pheromonal regulation of reproduction will be necessary to begin to untangle the complex factors involved in differential reproduction by nestmate queens in polygynous fire ants.

CONCLUSIONS

This brief overview reveals an unsuspected level of complexity in the reproductive and social structure of polygynous

TABLE 3. Characteristics of queens exhibiting weight-loss syndrome associated with high representation of their progeny among female reproductives (Source: Ross 1988)

Study Colony	Queen	Weight loss (mg)	Period of weight loss (days)	Queen death following weight loss?	Observed proportion of sexuals derived from queen during period of weight loss (expected[a])
COL C2	C54	21.4 → 9.4	88	yes	0.56 (0.25)
	C43[b]	16.7 → 8.2	34	no	0.84 (0.33)
COL D1	D53	19.4 → 7.8	66	yes	0.69 (0.17)
	D33	19.7 → 8.6	72	yes	0.84 (0.20)
COL J1	J54	20.1 → 10.6	52	yes	0.70 (0.20)
COL R1	R37	15.3 → 9.7	45	yes	0.90 (0.17)
	R5	19.0 → 8.6	30	yes	0.88 (0.25)
COL R2	R56[b]	21.2 → 8.7	32	yes	0.35 (0.25)

[a]Represents mean proportion of sexuals derived from each mother queen in the colony during this period
[b]Diploid-male-producing queens

S. invicta. The reproductive potential of dealate queens in polygynous colonies differs dramatically among the three reproductive classes of queens, and variability in individual reproductive competency is superimposed on this fundamental distinction of reproductive roles. It is reasonable to suspect that similar complexity in reproductive function may characterize other polygynous societies in the social insects. Accurate description of such complexity likely will depend on integrated physiological, behavioral, and genetic approaches.

Because of the importance of understanding reproductive roles in order to correctly identify the evolutionary forces which mediate colony social organization, a cataloguing of reproductive and social structures in a diversity of taxa exhibiting polygyny would seem to be requisite to further sustained theoretical advance. In particular, the importance of kin selection as a factor in the emergence and maintenance of polygyny may be assessed by determining inclusive fitness effects associated with the formation of multiple-queen groups, which determination requires knowledge of patterns of relatedness and individual reproductive roles. The tools are presently at hand to begin to fill these crucial gaps in our understanding of insect sociality.

Acknowledgments

I thank E. Vargo, D. Fletcher, and M. A. Moran for access to unpublished data and helpful comments. T. Brooks, S. Bradley, S. Hall, K. Hiett, and J. Robertson provided valuable technical assistance. The research presented here was funded in part by NSF grants PCM-8209097 (M. Blum and D. Fletcher) and BSR-8615238 (K. Ross) and grants from the American Farm Bureau Federation.

LITERATURE CITED

Adams, J., E. D. Rothman, W. E. Kerr and Z. L. Paulino. 1977. Estimation of the number of sex alleles and queen matings from diploid male frequencies in a population of *Apis mellifera*. *Genetics* 86:583-596.

Brian, M. V. 1983. *Social insects, ecology and behavioural biology.* Chapman and Hall:London.

Crozier, R. H. 1977. Evolutionary genetics of the Hymenoptera. *Ann. Rev. Entomol.* 22:263-288.

Crozier, R. H. 1979. Genetics of sociality, pp. 223-286. In *Social insects, vol I,* H. R. Hermann, ed. Academic Press:New York.

Evesham, E. J. M. 1984. Queen distribution movements and interactions in a semi-natural nest of the ant *Myrmica rubra* L. *Insectes Sociaux* 31:5-19.

Fletcher, D. J. C. 1983. Three newly-discovered polygynous populations of the fire ant, *Solenopsis invicta*, and their significance. *J. Georgia Entomol. Soc.* 18:538-543.

Fletcher, D. J. C. 1986. Triple action of queen pheromones in the regulation of reproduction in fire ant (*Solenopsis invicta*) colonies. *Adv. Invert. Reprod.* 4:305-316.

Fletcher, D. J. C. and M. S. Blum. 1983a. Regulation of queen number by workers in colonies of social insects. *Science* 219:312-314.

Fletcher, D. J. C. and M. S. Blum. 1983b. The inhibitory pheromone of queen fire ants: effects of disinhibition on dealation and oviposition by virgin queens. *J. Comp. Physiol.* 153:467-475.

Fletcher, D. J. C., M. S. Blum, T. V. Whitt and N. Temple. 1980.

Monogyny and polygyny in the fire ant, *Solenopsis invicta.*
Ann. Entomol. Soc. Amer. 73:658-661.

Fletcher, D. J. C. and K. G. Ross. 1985. Regulation of reproduction in eusocial Hymenoptera. *Ann. Rev. Entomol.* 30:319-343.

Greenberg, L., D. J. C. Fletcher and S. B. Vinson. 1985. Differences in worker size and mound distribution in monogynous and polygynous colonies of the fire ant, *Solenopsis invicta* Buren. *J. Kansas Entomol. Soc.* 58:9-18.

Heinze, J. and A. Buschinger. 1987. Queen polymorphism in a non-parasitic *Leptothorax* species (Hymenoptera,Formicidae). *Insectes Sociaux* 34:28-43.

Hölldobler, B. and N. F. Carlin. 1985. Colony founding, queen dominance and oligogyny in the Australian meat ant *Iridomyrmex purpureus. Behav. Ecol. Sociobiol.* 18:45-58.

Hölldobler, B. and E. O. Wilson. 1977. The number of queens: an important trait in ant evolution. *Naturwissenschaften* 64:8-15.

Hölldobler, B. and E. O. Wilson. 1983. Queen control in colonies of weaver ants (Hymenoptera:Formicidae). *Ann. Entomol. Soc. Amer.* 76:235-238.

Hung, A. C. F., S. B. Vinson and J. W. Summerlin. 1974. Male sterility in the red imported fire ant, *Solenopsis invicta. Ann. Entomol. Soc. Amer.* 67:909-912.

Jeanne, R. L. 1980. Evolution of social behavior in the Vespidae. *Ann. Rev. Entomol.* 25:371-396.

Mercier, B., L. Passera and J.-P. Suzzoni. 1985. Étude de la polygynie chez la fourmi *Plagiolepis pygmaea* Latr. (Hym. Formicidae) I. La fécondité des reines en condition expérimentale monogyne. *Insectes Sociaux* 32:335-348.

Michener, C. D. 1964. Reproductive efficiency in relation to colony size in hymenopterous societies. *Insectes Sociaux* 11:317-341.

Nonacs, P. 1988. Queen number in colonies of social Hymenoptera as a kin-selected adaptation. *Evolution* 42:566-580.

Pamilo, P. 1982. Genetic population structure in polygynous *Formica* ants. *Heredity* 48:95-106.

Pamilo, P. and R. Rosengren. 1984. Evolution of nesting strategies of ants: genetic evidence from different population types of *Formica* ants. *Biol. J. Linn. Soc.* 21:331-348.

Plateaux, L. 1981. Difficulté du remplacement de la reine dans une colonie de la fourmi *Leptothorax nylanderi. Ann. Sci. Nat. Zool. Paris* 3:1-14.

Ross, K. G. 1988. Differential reproduction in multiple-queen colonies of the fire ant, *Solenopsis invicta* (Hymenoptera:Formicidae). *Behav. Ecol. Sociobiol.* in press.

Ross, K. G. and D. J. C. Fletcher. 1985a. Comparative study of genetic and social structure in two forms of the fire ant *Solenopsis invicta* (Hymenoptera:Formicidae). *Behav. Ecol. Sociobiol.* 17:349-356.

Ross, K. G. and D. J. C. Fletcher. 1985b. Genetic origin of male diploidy in the fire ant, *Solenopsis invicta* (Hymenoptera: Formicidae), and its evolutionary significance. *Evolution* 39:888-903.

Ross, K. G. and D. J. C. Fletcher. 1986. Diploid male production - a significant colony mortality factor in the fire ant, *Solenopsis invicta. Behav. Ecol. Sociobiol.* 19:283-191.

Ross, K. G., E. L. Vargo and D. J. C. Fletcher. 1988. Colony genetic structure and queen mating frequency in fire ants of the subgenus *Solenopsis* (Hymenoptera:Formicidae). *Biol. J. Linn. Soc.* 34:105-117.

Thorne, B. L. 1984. Polygyny in the neotropical termite *Nasutitermes corniger*: life history consequences of queen mutualism. *Behav. Ecol. Sociobiol.* 14:117-136.

Vargo, E. L. and D. J. C. Fletcher. 1986. Evidence of pheromonal queen control over the production of male and female sexuals in the fire ant, *Solenopsis invicta. J. Comp. Physiol. A* 159:741-749.

Vargo, E. L. and D. J. C. Fletcher. 1987. Effect of queen number on the production of sexuals in natural populations of the fire ant, *Solenopsis invicta. Physiol. Entomol.* 12:109-116.

Vargo, E. L. and D. J. C. Fletcher. 1988. On the relationship between queen number and fecundity in polygyne colonies of the fire ant *Solenopsis invicta. Physiol. Entomol.* in press.

Voss, S. H. 1985. Rapid, simple DNA staining for fire ant eggs. *J. Entomol. Sci.* 20:47-49.

Voss, S. H. J. F. McDonald and C. H. Keith. 1988. Production and abortive development of fire ant trophic eggs. In *Advances in Myrmecology*, J. C. Trager, ed. Brill:New York. pp. 517-534.

West-Eberhard, M. J. 1981. Intragroup selection and the evolution of insect societies. In *Natural selection and social behavior*, R. D. Alexander and D. W. Tinkle, eds. Chiron Press:New York pp. 3-17.

West-Eberhard, M. J. 1983. Sexual selection, social competition, and speciation. *Quart. Rev. Biol.* 58:155-183.

Wheeler, D. E. 1986. Developmental and physiological determinants of caste in social Hymenoptera: evolutionary implications. *Am. Nat.* 128:13-34.

Wilson, E. O. 1971. *The insect societies.* Belknap Press of Harvard Univ. Press:Cambridge, Mass.

Yamauchi, K., K. Kinomura, and S. Miyake. 1982. Sociobiological studies of the polygynic ant *Lasius sakagamii.* II. Production of colony members. *Insectes Sociaux* 29:164-174.

CHAPTER 9

INCOMPATIBILITY BETWEEN COLONIES OF ACACIA ANTS: GENETIC MODELS AND EXPERIMENTAL RESULTS

Alex Mintzer

Models in general and computer-based models in particular force the investigator to explicitly state the underlying ideas and assumptions. Here, several explicit models of nestmate recognition are developed, each of which predicts patterns of incompatibility between colonies. The models concern genetic and physiological mechanisms for the generation of olfactory differences and learned criteria for assessing relationships of different individuals. The predictions of each model are examined in detail and experimental results are discussed in this context in an attempt to generate testable hypotheses about the odor production and discrimination mechanisms underlying nestmate recognition.

Although these models are applied specifically to my results on *Pseudomyrmex*, they are equally applicable to the analysis of nestmate recognition and colony odor in other social insects. The simple procedures developed here for examining and evaluating data expressed in the form of a compatibility table or matrix may find more general use as systematic data on pairwise recognition and compatibility patterns become more extensive in animal behavior and immunobiology.

INTERCOLONY INCOMPATIBILITY AND DISCRIMINATION OF NON-NESTMATES IN THE ANT *PSEUDOMYRMEX FERRUGINEA*

Pseudomyrmex ferruginea is a well known obligate ant inhabitant of swollen thorn acacias of lowland Central America.

Department of Entomology, Texas A&M University, College Station, TX, USA. Current address: 1550 S. Marmora, Tucson, AZ 85713 USA

Their brood is reared in the large hollow thorns of the acacia, and the ants subsist on foliar nectar and small detachable plant-produced structures termed Beltian bodies, which are rich in lipid and proteins required for larval growth. As Janzen (1966) demonstrated, colonies occupy entire shoots and defend them against intruding herbivores and foreign conspecifics, by using a potent sting as their principal weapon. The discrimination of non-nestmates is an olfactory process, involving receptors on the antennae. Visual perception of behavioral cues is not involved, as acacia ants can readily distinguish immobile nestmates from non-nestmates in the dark.

Experimental Compatibility Table

Twenty-one colonies were reared from founding queens on a clonal population of the host plant *Acacia hindsii* in the greenhouse. Under these conditions the plant-produced diet and nesting environment of the colonies is standard and uniform. However, the ants remain capable of distinguishing nestmates from non-nestmates, indicating that the ants themselves are the likely source of the chemical odor cues or labels used. Extensive pairwise compatibility testing generated the results shown in TABLE 1 for 19 greenhouse colonies. Foraging ants were collected from donor colonies and introduced on the exterior surface of short test branches taken from recipient colonies. Most non-nestmates were blocked or attacked at the thorn entrance, or ejected within 24 h after obtaining initial entry. Resident ants and approximately 30% of introduced non-nestmates were accepted, entering the thorns and remaining inside unchallenged. Rejected ants were frequently injured or killed. Pairwise compatibility testing showed that rejections occurred among most combinations in the compatibility table, but a few pairings gave only acceptances with four to six replications. To examine rejection frequencies in more detail, a 5x10 section of the compatibility table was selected. I conducted 20 compatibility tests for each paired combination of the colonies involved, with the results shown in TABLE 2. With additional replications, patterns of consistent acceptance disappear, but a range of rejection frequencies (30-100%) are still observed for the paired combinations. In this test series (n=900), the overall rejection frequency was 77%. Finally, 30 additional compatibility tests were completed for each of three paired combinations. These tests, conducted over a period of three years, showed that

Recipient Colony

Donor Colony

TABLE 1. Compatability matrix for the 19 experimental colonies of *Pseudomyrmex ferruginea*. X indicates that both introduced ants were rejected. O indicates that both or all introduced ants were accepted. / indicates introduced ants were both accepted and rejected. Superscripts indicates number of replications, if other than two.

Recipient Colony	1	2	3	4	5	6	7	8	9	10	11	12	13	14	15	16	17	18	19
1	0	X	X	X	X	X	X	X	X	/	X	X	X	X	X	X	X	X	X
2	/⁴	O	/⁴	/³	O	X³	X	/	/	/	/	/	/	X	/	X	X	X	/
3	/³	/⁴	O	/	/⁴	/⁴	/	/	X	X	/	X	X	X	X	X	X	X	X
4	X	Ó⁴	X	O	X	X	X	X	X	X	X	X	X	X	X	X	X	/	X
5	X	X	/	O	O	X	X	/	X	X	X	.	.	.	.	.	.	.	.
6	/	X	/	/	/	O	/	O	/	/	O⁵	.	.	.	.	.	.	.	.
7	X	X	X	X	X	/³	O	X	X	X	X	X	X	X	X	X	X	X	X
8	X	/³	X	X	/¹	X	O⁵	O	X	O	/⁵	X	X	X	X	X	X	X	X
9	/	/⁴	/	/²	/¹	/⁷	/⁴	X	O	X¹	X	/	/	/	/	/	/	/	X
10	X	X³	/	X	X¹	X	X⁶	X	X¹	O	O	O	O	X	O	O	X	/	X
11	X	/³	/³	/⁴	O¹	X	X⁴	X	/³	X	O	O⁵	X	X	X	X	/	X	/
12	X	X	X	X	.	/	X⁴	/	X	X²	X	O	X	X	X	X	X	X³	X
13	X	X	/	X	.	X	X	X	X	X	X	X	O	O	O	O	O	O	/
14	X	X	X	X	.	X	X	X	X	X	X	O	/⁴	/⁴	X	X³	X³	X⁴	X
15	X	/	X	X	.	X	X	X	X	X	X	O³	O³	/	O	O³	O	O	O³
16	Ó⁴	X	X	X	.	/	X	/³	X	O	/³	/³	/³	/	X	X	/	O	O³
17	X	X	X	X	.	X	X	/	X	/	O³	X	O³	X³	X	X³	O	/	/³
18	/³	/	X	X	.	X	X	X	X	X	X	/	O³	X³	/	X³	O	O	O³
19	X	X	X	X	.	X	X	X	X	X	X	/	/	X	X	O³	X	O³	O

TABLE 2. Number of rejections in 20 replicate trials for each paired combination of source and recipient colonies.

Recipient Colony	Source Colony				
	NAM2	ES2	LL2	LL3	NAM5
NAM2	- -	14	13	14	15
ES2	12	- -	6	9	12
LL2	19	12	- -	19	17
LL3	17	14	15	- -	19
NAM5	19	20	20	19	- -
NAM1	18	17	19	17	19
NAM3	15	19	15	20	19
NAM4	18	18	20	19	9
NM5	13	6	7	11	18
SM2	19	15	14	10	19

rejection frequencies for specific paired combinations are constant over a long period and differ significantly from frequencies for other paired combinations.

Factors Affecting the Experimental Compatibility Table

Although the tests were conducted under standardized conditions, two problematic influences could not be avoided completely. First, all residents do not react to an introduced ant in the same manner. Young workers in particular were unaggressive towards nonresidents, while the older foraging workers reacted very strongly to conspecifics and disturbances in general. Some test branches occupied by young workers only may have accepted nonresidents which would have been rejected by older workers. 'False negative' reactions of this type would reduce the observed overall rejection frequency, and obscure the occurrence of uniform rejection outcomes (x) in the compatibility table. Second, some introduced ants may provoke the fight with the residents, in cases in which the residents would have accepted a nonaggressive nonnestmate. In over 1000 tests, I observed this fewer than five times on the thorn exterior, when the introduced ant lunged and seized a resident at the thorn entrance. I have no observations as to how often this occurred inside the thorn. Such 'false positives' increase the observed rejection frequency. Fortunately, the aggressive patrolling ants were readily distinguished and avoided when test ants were collected.

Experimental Inbreeding and Incompatibility

The entire acacia ant colony life cycle can occur in the greenhouse. Colonies begin producing daughter queens and males when they are one to two years old, and this production of winged ants continues thereafter without interruption. Daughter queens and males were collected from two parent colonies and allowed to mate in closed containers, providing a start for three inbred lineages and other outcrossed lineages. Mated queens were provided with vacant thorns on unoccupied acacia plants. Many of these ants successfully established colonies and the experimental breeding process was repeated when these daughter colonies reached maturity. Pairwise compatibility tests were conducted between inbred daughter colonies, and between daughter colonies and the original colonies of the parent generation. A modest but significant decrease in overall non-nestmate rejection frequencies was observed within the three inbred lineages, compared with the rejection frequency of the original colonies. The rejection frequency dropped from 77.7% for unrelated parental colonies to 68% (n=120), 62.5% (n=200), and 55% (n=20) in inbred F_1 lineages. No additional decrease in rejection frequency was noted in compatibility tests between F_2-generation inbred colonies in two lineages (65% of 20 replications, and 70% of 50 replications).

SPECIFIC CONSIDERATIONS OF THE MODELS

Recognition pheromones

Environmental heterogeneity may be an unreliable source of colony odor cues, even when it has the potential to produce significant differences or variation. Neighboring colonies cannot expect environmental differences to provide adequate odor label separation, nor can acacia ant colonies be expected to modify their diets in order to enhance intercolony odor differences.

Environmental odor effects are an extrinsic stochastic influence and are very difficult to quantify in models like the ones described here. Even though these influences are not incorporated in the models, their effects can be important in local situations for some species (Obin 1986).

Genetic Models

Specific models were developed for two extreme cases: (1) Odor determined by a single-locus system with multiple alternative alleles, and (2) Odor determined by a multiple- locus system with two alternative alleles at each locus. Two independent karyotype examinations indicate that *P. ferruginea* has a haploid chromosome number n~20 (W.L. Brown personal communication, J.S. Johnston personal communication).

In most cases, alternative alleles were assumed to be equally frequent in the source population. This situation is easiest to model, and it is also likely to result from a frequency-dependent selection process. However, one model variation incorporated a multilocus system with one rare allele ($p=0.1$) and a common alternative ($q=0.9$) at each locus.

In heterozygotes, alternative alleles were assumed codominant, with both allelic products expressed in the phenotype. No secondary interaction, compounding, or synergistic combinations of allelic odor components were considered; each allelic product is a distinct element. The olfactory distinctions are based on qualitative differences between individuals or colonies, not quantitative differences such as pheromone concentration.

Caste Source of Pheromones

Male ants can be eliminated as a source of recognition cues, because they are absent in young colonies and are only seasonally present in mature ones. The development of nestmate recognition capabilities shows no corresponding pattern.

If the queen is the pheromone source in the monogynous colony, a true colony odor would result as recognition pheromones from a single source are spread by grooming and/ or oral food exchange between individuals. The odor label on each worker should be uniform in composition, but it might vary in concentration or intensity depending on distribution patterns.

Patterns of allogrooming and food exchange could assume great importance if the workers or brood are the primary source of recognition pheromones. Many different individuals would be producing pheromones; colony members are genetically related but not identical, and this should be expressed in phenotypic variation in individual odor contributions. The extent to which individual contributions are mixed and blended (as opposed to remaining distinct and separate) in the colony is crucial. Crozier and Dix (1979) also raise this question in their theoretical modeling attempts with colony-specific recognition. Mixing could also result if secretions containing recognition pheromones are voided or smeared inside the nest and the cuticle absorbs odoriferous components from marked nest surroundings. Complete blending of individual contributions would produce a 'gestalt odor'.

Factors Influencing Genetic Relationships

In most species of ants, queens probably mate several times. This is certainly true in *Pogonomyrmex*, but Janzen (1973) reports only single matings in *Pseudomyrmex ferruginea*. I have observed over 30 females copulating in small containers, and two females did copulate twice (in one case, both times with the same male). Additional female copulations will decrease the average relatedness between workers and increase the potential number of alternative alleles incorporated in each colony.

Colonies with several functional queens will also show decreased average relatedness between workers and increased genetic diversity within each colony. Colonies of *P. ferruginea* are founded by a single female, and no additional females are adopted. I have developed the models for monogynous colonies, founded by queens that have mated once and twice.

Ontogeny of Recognition Behavior: Learned Criteria

As Alexander and Borgia (1978) and others point out, a genetic model of recognition completely excluding learning is difficult to visualize. All of the models presented here assume a period of social learning early in the adult stage, leading to the establishment of a set of olfactory criteria (template) used in phenotypic comparisons with other ants. One consideration here is which individuals are used in developing the criteria. In the social insects, the criteria are established within the parent colony, before any non-nestmates are encountered. Buckle and Greenberg (1981) showed that an individual of the primitively eusocial bee *Lasioglossum zephyrum* does not use its own odor contributions in developing olfactory criteria, but does use its nestmates' odor. The models described here were developed for ants in colonies with many (>100) nestmates to examine, and

169

assume that ants do not directly assess their own odor phenotype. They also assume a dichotomy in behavioral response, based on a 'cue similarity threshold' as proposed by Gamboa *et al.* (1986).

During the learning period, ants could be attempting to identify 'common denominator' components in all individuals in the colony. In this model, ants learn to identify allelic products shared by all colony members. Individuals lacking the label component are considered non-nestmates. Crozier and Dix used this approach in their genetic models, and they recognized that it works best if the colony queen mates with one male. The criteria described here are based on the 'greatest common denominator'. When a common denominator criterion is developed for a colony comprised of identical heterozygotes, both allelic components are required simultaneously.

As an alternative, ants might identify familiar allelic components in all colony members. In this model, ants habituate to all odor components encountered during the learning period; thereafter, individuals with unfamiliar label components are considered non-nestmates and rejected. For my simulations, I used both learning criteria models, hereafter referred to as the 'common denominator' and 'foreign label rejection' models (adopting some terminology of Getz 1982).

PROCEDURE FOR MODEL SIMULATIONS AND ANALYSIS

Generation of Genotypes and Compatibility Matrix

Single-locus simulations were run with 2-10 alternative alleles. For each specific number of alleles, I ran ten simulation runs with each model variation. Multilocus simulations incorporated two to nine independently assorting gene loci, except that ten loci were used in some 'rare allele' model simulations.

After the number of gene loci and alternative alleles was specified, a uniformly distributed random number generator was used to assign genotypes to 20 diploid females and 20 haploid males. A second set of 20 haploid genotypes was generated when double insemination of females was modeled. These 'individuals' constituted 20 'mated pairs',and the genotypes of diploid 'worker offspring of each pair were generated. If a gestalt colony odor was specified, each aggregate parental genotype (for singly inseminated females, a triploid) was retained for all workers in the colony, and the segregating offspring genotypes were not used.

Given a specific learning model, nestmate recognition

criteria were developed for each 'colony' and the 'worker' genotypes from every other colony were compared and evaluated. This generated a compatibility table (see TABLE 3 for an example). In all cases, a diagonal of zeroes on the compatibility table indicates nestmate recognition and acceptance for each 'colony' in the model output. Thus, the model compatibility tables have the same format as the experimental table.

Analysis of Model Output

The number of homozygous colonies [e.g. (aa)x (a)] and heterogeneous colonies [e.g. (ab)x (c)] generated in each simulation run was recorded.

The overall non-nestmate rejection frequency was calculated for each compatibility table. When the overall table rejection frequency was close to the 95 percent confidence interval for the experimental result, I conducted additional analyses. Individual row and column totals were derived for each colony in the table. From these values, two non-nestmate rejection frequencies can be calculated for each colony; the first value reflects performance as a recipient of introduced ants, while the second measures performance as a source of ants introduced into other 'colonies'. I prepared a scatter-plot of the two measures (one point for each colony) in the simulation output.

Finally, the symmetry of each compatibility matrix across the diagonal was examined. Table elements on the diagonal were excluded from the analysis. The other elements $T(i,j)$ were compared with their mirror images $T(j,i)$ across the diagonal.

RESULTS AND DISCUSSION

Compatibility Table Elements

Using the 'gestalt' model, all members of a colony share the same odor phenotype. Consequently, all members from a given colony meet the same fate in paired compatibility tests, and the table elements would reflect only two outcomes (x and o). This predicted result does not compare favorably with the experimental results. In many cases, both acceptance and rejection were observed in replications of a given pairing of colonies. Under the gestalt model, a minor occurrence of such mixed outcomes might be ascribed to failure of young residents to reject non-nestmates in some replications. However, the widespread occurrence of

171

Table 3. Simulation output for a single locus model incorporating the "common denominator" learned criteria.

GENOTYPES				COMPATIBILITY MATRIX																		
		Worker																				
Male	Female	Offspring		1	2	3	4	5	6	7	8	9	10	11	12	13	14	15	16	17	18	19
b	ce	bc	be	1	O	O	~	~	~	O	O	~	X	O	~	O	O	X	X	~	~	X	X
b	eb	be	bb	2	O	O	~	~	~	O	O	~	X	O	~	O	O	X	X	~	~	X	X
e	cb	ec	eb	3	~	~	O	X	X	~	X	X	O	X	X	X	X	~	O	O	X	X	O
c	bd	cb	cd	4	~	X	~	O	O	X	X	X	X	X	O	X	X	~	~	X	O	X	X
c	cb	cc	cb	5	~	X	~	O	O	X	X	X	X	X	O	X	O	~	~	X	O	X	X
b	de	bd	be	6	O	O	~	~	~	O	O	~	X	O	~	O	~	X	X	~	~	X	X
d	bb	db	db	7	X	X	X	X	X	O	O	O	X	X	X	O	~	X	X	X	X	X	X
d	db	dd	db	8	X	X	X	~	X	~	O	O	X	X	X	O	~	O	X	~	X	O	O
e	aa	ea	ea	9	~	~	X	X	X	X	X	X	O	X	X	X	X	~	~	X	X	X	X
b	bb	bb	bb	10	O	O	~	X	X	O	O	~	X	O	~	O	O	X	X	~	~	X	X
c	cb	cc	cb	11	~	X	~	O	O	X	X	X	X	X	O	X	X	~	X	X	O	X	X
b	dd	bd	bd	12	O	O	X	X	X	O	O	O	X	O	X	O	O	X	X	O	X	X	X
b	db	bd	bb	13	O	O	X	~	O	~	O	O	X	O	~	O	O	~	X	X	~	X	X
d	ce	dc	dc	14	X	X	~	~	X	X	~	X	X	X	X	~	~	O	~	X	X	~	X
e	cd	ec	ed	15	X	X	X	X	X	X	O	~	O	X	X	X	X	O	O	X	X	O	O
e	bd	ab	ed	16	~	~	O	X	X	~	X	X	O	X	X	O	X	X	O	O	X	X	O
c	ba	cb	ca	17	~	~	O	O	O	~	X	X	X	X	O	X	X	X	O	O	X	X	O
a	dd	ad	ad	18	X	X	X	X	X	X	X	X	X	X	X	X	X	~	X	X	X	O	X
d	ee	de	de	19	X	X	X	X	X	X	X	X	X	X	X	X	X	X	~	~	X	X	O

mixed outcomes in the experimental results makes any explanation based on the gestalt model difficult to sustain.

With other single locus models, two segregating worker genotypes and phenotypes are recognized in each colony, and the compatibility table model will show three possible outcomes (see TABLE 3). The third outcome (/) indicates a cell rejection frequency p=0.5, where only one worker segregant phenotype is accepted. Colonies founded by females that mate twice show a greater variety of worker genotypes than colonies originating from singly-mated females. Up to four segregant genotypes are recognized per colony, and two intermediate rejection frequencies are noted. Two of the four possible genotypes may be accepted, yielding a cell rejection frequency p=0.5, or only one of the four may be accepted, yielding a rejection frequency p=0.75.

The situation is also more complicated with multi-locus models. In addition to the uniform outcomes (x and o), several intermediate rejection frequencies are possible. The most common outcome reflects a rejection frequency p=0.5, where compatibility is determined by segregants at one locus. Other higher rejection frequencies (0.75,0.875, 0.938 respectively) result from cases determined by segregants at several loci. In the table, these higher rejection frequencies occur less often that p=0.5, with p=0.938 least common of all.

In model simulations, the individual table elements represent definitive expected outcomes and are independent of sample size and errors. However, the elements in the first experimental table were derived from a very limited series of replications (n=1 to 7). This severe sample size limitation forces some caution in the interpretation of the initial experimental results, particularly at the level of individual table elements. With additional sampling, an experimental $\underline{O}$ or $\underline{X}$ may turn into a $\underline{L}$, but the reverse will never happen. Intensive resampling of a 5x10 section of the original experimental table showed that the initial results underestimated the occurrence of mixed outcomes in the table. FIGURE 1 shows the distribution of rejection frequencies for the 45 paired colony combinations in TABLE 2. The abundance of intermediate cell rejection frequencies in experimental results supports individual odor label models in preference to gestalt alternatives, but indicates no clumping of observed rejection frequencies around the predicted values p=0.5 or 0.75.

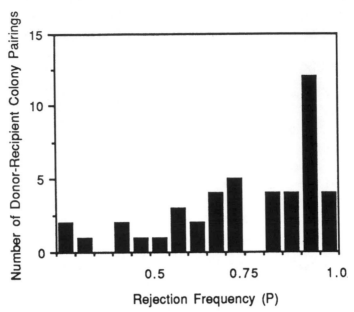

FIGURE 1. Distribution of rejection frequencies for the experimental results in TABLE 2.

Overall Frequency of Rejection

The rejection frequency increased as the number of alternative alleles or independently assorting gene loci was increased. Rejection frequencies which approximated the experimental result could be obtained using appropriately selected parameters with all model variations except the single locus gestalt model for colonies founded by singly inseminated females.

For single-locus multiple-allele models the rejection frequency increased most rapidly for the 'gestalt model'; the other two models had rejection frequencies which approached the experimental results more closely (FIGURE 2A). The 'foreign-label-rejection' learned criteria model best approximated the actual results in simulations with five alleles, while the common denominator learned criteria model best approximated the experimental rejection frequency with five to eight alleles. With more than eight alternative alleles, all three model variations produced rejection frequencies which exceeded the experimental rejection frequency.

Although the pattern is similar for the single-

174

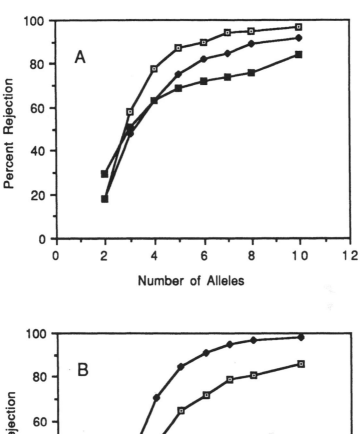

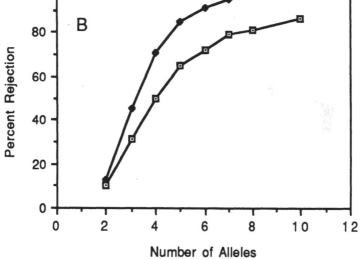

FIGURE 2. Rejection frequency as a function of the number of alternate alleles for single locus models. A. Single-insemination case. Upper line with empty squares--gestalt model. Line with diamonds--foreign label rejection model. Line with filled squares--common denominator model. B. Double insemination case. Upper line--gestalt model. Lower line--foreign label rejection model.

insemination and double- insemination variations of the models, the frequency of rejection is lower at each allele number level for the two-insemination models (FIGURE 2B). The experimental result was approximated with gestalt model simulations with four alleles and 'foreign-label-rejection' model simulations with five to seven alleles.

Overall rejection frequencies for multilocus model simulations approximated the experimental results with simulation models assuming five to seven independently assorting gene loci for 'foreign-label-rejection' learned criteria and three loci for 'common denominator' learned criteria. For the 'rare allele' model using the common denominator criteria, simulation output with nine independent loci best approximated the experimental rejection frequency.

Colony Rejection Frequencies

FIGURE 3 is a scatter plot of the rejection frequencies for the 19 experimental colonies. FIGURE 4 shows scatter plots of individual colony rejection frequencies for representative simulation output. In the figures, each model output produces a nebulous cloud of points without a tight correlation. However, the orientation of the point distributions is visibly different for the two learned criteria models.

If the ants discriminate non-nestmates by the presence of unfamiliar components not encountered on nestmates, the behavior of homogeneous (i.e. homozygous) colonies and heterogeneous colonies should differ. The 'foreign-label-rejection' criteria model predicts that homogeneous colonies have high rejection frequencies when considered as recipients and lower rejection frequencies when considered as sources of ants introduced into other colonies. Homogeneous colonies would reject more frequently because the criteria of such colonies are narrowly based and they encounter unfamiliar allelic components frequently. However, this 'narrow base' should render workers from homogeneous colonies more acceptable on the average to other colonies than workers from heterogeneous colonies. Heterozygous workers are more likely to incorporate a foreign allelic odor component which would lead to rejection. On the scatter plot, homogeneous colonies should be concentrated above and to the left of the heterogeneous colonies. As the frequency of homozygous colonies decreases with increasing allele numbers, this relationship will be obscured.

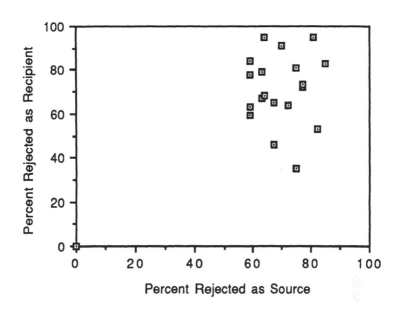

FIGURE 3. Individual experimental colony rejection frequencies, showing performance as donor and recipient of introduced ants.

The 'common denominator' criteria model predicts that homozygous workers will be accepted less often than heterozygotes, since the latter incorporate more alternative alleles, one of which might lead to acceptance. In most cases, homogeneous and heterogeneous colonies will be using only one allele to identify nestmates, so their rejection frequencies as recipients may not differ significantly. The exceptions are colonies comprised exclusively of identical heterozygotes, which will use both allelic components simultaneously and only accept introduced individuals with identical genotypes.

A marked contrast in predictions occurs with the multi-locus, two allele models. Some heterogeneous colonies incorporate both alleles at every odor locus, and thus lose any capacity to discriminate non-nestmates on the basis of unfamiliar components (there being none!). These colonies appear as a group of points on the x-axis in FIGURE 4A. These become predictably less frequent as the number of loci increases, but they still occur in almost every simulation when the model is run with seven to

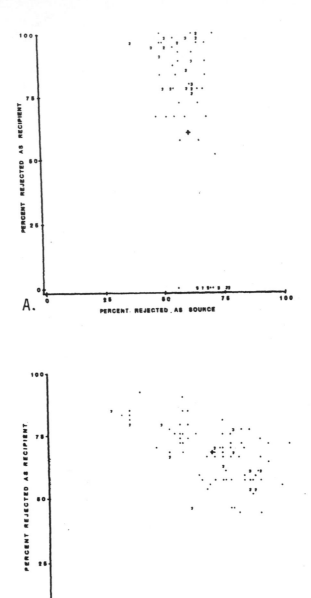

FIGURE 4. Individual model colony rejection frequencies, showing performance as donor and recipient of introduced ants using multi-locus diallelic models. Each distribution was generated with pooled data from five simulation runs. A. Foreign label rejection model. B. Common denominator model. + indicates center (mean rejection frequency) of each distribution.

nine loci. The 'common denominator' learned criteria model makes no similar prediction. Colonies comprised exclusively of heterozygous workers (AB/CD/EF...) will be most selective, accepting only similar heterozygous workers in turn. Workers from such colonies will be accepted by all other colonies. On the scatter-plot, such colonies would generate points along the y-axis. However, such colonies appear very infrequently when the model is run with four or more loci, where the overall rejection frequency approximates the experimental frequency. There are none in FIGURE 4B.

These two measures of rejection frequency are significant from an evolutionary perspective. Colonies that reject non-nestmates more frequently are less vulnerable to parasitism by invading egg-laying workers and queen destruction by unchallenged intruders. Conversely, colonies whose workers find widespread acceptance in the population may have a considerable selective advantage, if opportunities for worker oviposition are common. Colonies that combine these two features (i.e. those in the upper left corner of FIGURE 4) should have the greatest advantage. Over time, selective pressure can be visualized as driving the observed rejection values upwards on the figure and also towards the left, although the latter is obviously subject to a frequency dependent check. From this perspective, the multi-locus two-allele model using the 'foreign-label-rejection' criteria is especially unstable.

Compatibility Table Symmetry

None of the models described here generates a perfectly symmetrical compatibility table across the diagonal; all produce some asymmetrical outcomes. The matrix is less symmetric at intermediate overall rejection frequencies than at very high or low rejection frequencies.

At overall rejection frequencies approximating the experimental result, model output incorporating the 'foreign-label-rejection' criteria is less symmetrical. Due to the effect of the heterogeneous colonies discussed above the most asymmetrical tables were encountered in simulation output of multiple-locus models incorporating the 'foreign-label-rejection' learned criteria. Because these colonies accept all comers, a row of zeroes is generated for each of them, which decreases the overall symmetry across the diagonal.

When the more detailed rejection-frequency results in the

top half of TABLE 2 are compared across the diagonal, the mean difference in observed p values (p=p_{ij}-p_{ji}) between 20 paired, symmetrical cells is 0.205. This is close to the mean difference in observed p values across the diagonal for the 90 possible nonsymmetrical cell pairings (p=0.228).

Inbreeding Effects on Rejection Frequencies

Inbreeding reduces allelic diversity in the resulting lineage. This has an immediate and dramatic effect on single locus, multiple-allele systems. With a single sibmating, a maximum of three alternative alleles will be carried forward in the inbred offspring of the female. At best, the overall rejection frequency in an inbred F_1 lineage will approach 50% for either 'common denominator' or 'foreign-label-rejection' models (see Fig. 2A for 3 alleles). Because many matings carry only two alleles forward in the offspring, the overall rejection frequency will probably be significantly less than 50%. In contrast, homozygosity leading to reduced rejection frequencies would be achieved very slowly by inbreeding if many independent loci are involved.

Empirical results provide little support for single locus, multiple-allele models. The modest observed decrease in rejection frequencies within lineages founded by inbred sister queens is much more consistent with polygenic, diallelic models.

CONCLUSIONS

Single-locus, multi-allele models for nestmate recognition labels are not supported by empirical evidence in *Pseudomyrmex ferruginea*. On theoretical grounds, Crozier (1986) argues that such gene systems cannot be stabilized by selection for recognition labels. Instead, available data support more traditional and complex polygenic, probably diallelic, models for recognition pheromone components. Quantitative factors such as differences in component concentration may have a larger role than unique component identity. Nestmate recognition labels can be removed from the epicuticle of acacia ants with nonpolar solvents. Analysis of nonpolar extracts show that ants from different colonies do not display any unique colony-specific components among the formidable mix of more than 30 detected cuticular hydrocarbons. Details of more complex quantitative genetic models required to generate such label component patterns remain unclear, but the occurrence of repetitive component patterns may simplify the task ahead. Chromosome numbers influence the number of potential independent loci available for a recognition label system. It would be interesting to examine the chemistry, ecological context, and genetics of nestmate recognition labels in ant species with low chromosome numbers. Progress on the ontogeny and sensory physiology involved in this olfactory discrimination is needed to complete the picture. A satisfying understanding of olfactory capacities and criteria or templates used in discrimination of nestmates requires collaboration with neurobiologists, to clarify the information pathways and processing from the antennal receptors through interneurons to the brain.

Acknowledgements

The models described here were originally developed during Ph.D. thesis research at the University of Michigan, using BASIC language on a Commodore PET microcomputer. Work on inbreeding of acacia ants at Texas A&M University was supported by the National Science Foundation (BNS-8303402) and by the Texas Agricultural Experiment Station.

LITERATURE CITED

Alexander, R. D. and G. Borgia. 1978. Group selection, altruism and the levels of organization of life. *Ann. Rev. Ecol. Syst.* 4:325-383

Crozier, R. H. 1986. Genetic aspects of kin recognition: Concepts, models and synthesis. In *Kin recognition in animals*, D. J. C. Fletcher and C. D Michener, eds. John Wiley:New York. pp. 55-73.

Crozier, R. H. and M. Dix. 1979. Analysis of two genetic models for the innate components of colony odor in social Hymenoptera. *Behav. Ecol. Sociobiol.* 4:217-224

Gamboa, G J., H. Reeve and D. Pfennig. 1986. The evolution and ontogeny of nestmate recognition in social wasps. *Ann. Rev. Entomol.* 31:431-454

Getz, W. M. 1982. An analysis of learned kin recognition in Hymenoptera. *J. Theor. Biol.* 99:585-597

Buckle, G. and L. Greenberg. 1981. Nestmate recognition in sweat bees (*Lasioglossum zephyrum*). Does an individual recognize its own odour or only odours of its nestmates? *Anim. Behav.* 29:801-809

Janzen, D. H. 1973. Evolution of polygynous obligate acacia-ants in western Mexico. *J. Anim. Ecol.* 42:727-740

Janzen, D. H. 1966. Coevolution of mutualism between ants and acacias in Central America. *Evolution* 20:249

Obin, M. 1986. Nestmate recognition cues in laboratory and field colonies of *Solenopsis invicta* (Buren). *J. Chem. Ecol.* 12:1965-1975

CHAPTER 10

EVOLUTIONARY GENETICS OF A PRIMITIVELY EUSOCIAL HALICTINE BEE, *DIALICTUS ZEPHYRUS*

Penelope F. Kukuk

The 2,000 or more species of Halictinae ("sweat bees") are a major component of the worldwide bee fauna (Sakagami 1980). The majority of known, primitively eusocial bee species are halictines, and it is an important taxon in the study of the origins and consequences of eusocial behavior in the Hymenoptera (Michener 1974, Wilson 1971). Most halictine bees construct nests by excavating narrow subterranean tunnels and small, individual brood chambers called *cells*, which sometimes are in clusters. Each cell is provisioned with a "loaf" of pollen mixed with nectar. A single egg is deposited upon each loaf. After oviposition the brood cell is closed, and little or no contact between adults and immatures occurs during development. In virtually all cases, eusociality in the halictine bees is primitive (but see Plateaux-Quénu 1962). The queen and workers are morphologically and physiologically similar, and queens can survive and reproduce alone. Moreover, all females are potential reproductives at emergence.

The reproductive division of labor of *D. zephyrus* , the most intensely studied halictine, is a result of dominance interactions among adult females (Brothers and Michener 1974, Buckle 1982a 1982b 1984 1985, Greenberg and Buckle 1981, Michener and Brothers 1974, Michener *et al.* 1971a 1971b). As all young females are physiologically and morphologically similar and behavioral interaction among adults is the primary agent of caste determination in *D. zephyrus,* this is an ideal species for the study of genetic mechanisms underlying maintenance of worker altruism.

Department of Entomology, Cornell University, Ithaca, NY, 14850, USA

D. zephyrus is a primitively eusocial species with overlapping generations, cooperative brood care, and reproductive division of labor. In New York, colonies are initiated in the spring by one, or occasionally two, females, one of which becomes a subordinate cofoundress (the caste definitions of Eickwort and Kukuk 1987, are used in this chapter). Successful single foundresses provision 2 to 7 brood cells and then become inactive while the brood matures. These initial cells primarily yield females that function as workers (or as replacement queens if the foundress dies). These small colonies (eusocial if the foundress queen survives and semisocial if she is replaced by one of her daughters) rear the reproductive brood of both males and gynes. Where the season is long enough for annual colonies to build up over more than one worker brood (e.g., in Kansas), a summer colony typically contains a queen, who is the primary egg producer, and 1 to 20 partially or fully nonreproductive workers (Batra 1966).

During colony ontogeny, dominance relationships emerge through interactions among young, adult females (Brothers and Michener 1974, Michener *et al.* 1971a 1971b, Buckle 1982a 1982b 1984 1985, Smith 1987). If the queen is removed, however, one of her workers soon replaces her (Breed 1977, Brothers and Michener 1974, Michener and Brothers 1974) and her workers are more likely to mate (Greenberg and Buckle 1981). Studies of brood rearing indicate that each cell is constructed and provisioned by more than one female but oviposition is, for the most part, the prerogative of the dominant female who is the functional queen (Batra 1964, Breed and Gamboa 1977, Kukuk and May in press).

POPULATION STRUCTURE

The evolutionary origin and maintenance of altruism requires the nonrandom association of genotypes. Nonrandom associations among genotypes occur at various levels of population structure and are due to either the mating structure or the association structure of a species (Wade 1980). The population structure of most social Hymenoptera is complex and multileveled as is the population structure of *D. zephyrus* (e.g., Berkelhammer 1982, Craig and Crozier 1979, Crozier 1977 1980, Crozier *et al.* 1984 1987, Halliday 1983, Lester and Selander 1981,

Metcalf 1980, Metcalf and Whitt 1977, Pamilo 1981 1982 1983 1984, Pamilo and Rosengren 1984, Pamilo and Varvio-Aho 1979, Pearson 1982 1983, Ward and Taylor 1981).

It is useful to view population structure in a hierarchical manner when investigating nonrandom associations among genotypes. This approach provides a useful framework and does not negate the importance of Hamilton's (1963 1964 1972) inclusive fitness model. Rather, this approach is consistent with recent treatments in which the distinction between kin selection and group selection has become somewhat blurred. This involves selection among groups whose members are genetically related (Fix 1985, Grafen 1986, Michod 1982, Michod and Abugov 1980, Nunney 1985, Taylor and Wilson 1988, Unoyama and Feldman 1980, Wade 1980, Wilson 1977 1980 1983 1987).

The population structure of *D. zephyrus* and other halictine bees (see Packer 1987) consists of hierarchical levels. Colonies of *D. zephyrus* occur in aggregations of a few to thousands of nests. Nests are found in patches of heavy soil in vertical, bare earth banks or rarely, in bare, horizontal soil (Batra 1966). Such banks are the result of disturbances and are often highly ephemeral. Floods can destroy aggregations of nests (observed in Kansas by Batra 1966, Crozier *et al.* 1987), or the *lack* of minor perturbation may allow early successional plants to proliferate and render the site unsuitable for *D. zephyrus* within 2 to 3 years (Kukuk personal observation).

Suitable habitat can thus become available for colonization or can become unsuitable rather rapidly. Local aggregations may be long or short lived, depending upon the frequency and intensity of disturbance at any given site and sites may change in their suitability over time. Along any single stream or river there is a linear patchwork of potential nest sites (i.e., heavy soil), some currently suitable (i.e., bare of plants) and others not (i.e., occupied by plants). In such a patchwork, local populations may become extinct as patches become unsuitable and, conversely, newly suitable habitat could be colonized to form new local populations.

The following hypothetical sequence of events could result in a complex, hierarchical structure in *D. zephyrus* populations: In the first year new habitat is colonized in the spring by one or a few inseminated females. A substantial proportion of these reproductives mate near their natal nests (both males and females) and the mated females overwinter in their natal nests. In the second year new colonies are initiated near the natal nests, thus creating neighborhoods of related colonies. This process is repeated as long as the habitat remains suitable. As the process continues, neighborhoods merge and secondary colonization by migrants from nearby aggregations cause the neighborhood structure within large aggregations to erode. After habitat change or population destruction this cycle repeats at the same or a new location.

Of particular interest is genetic variation at the levels of the aggregation, the neighborhood, the colony, and the individual. Data are available from studies in two regions, central New York and eastern Kansas. As Kansas is considerably farther south, *D. zephyrus* has a much longer active season there. The density of nest aggregations appears to be much lower in eastern Kansas than it is in central New York. In New York, I located and sampled 37 nest aggregations over a period of 6 weeks by driving along roads bordering creeks and stopping to examine apparently suitable habitat on foot. Such a task would have been impossible in eastern Kansas, where nest aggregations are found at greater distances from one another (Kukuk personal observation). Given these differences, it is not surprising that population structure and colony composition may differ between the two regions.

GENETIC VARIATION AMONG NEST AGGREGATIONS

There is conflicting evidence for genetic differentiation among adjacent aggregations. Crozier *et al.* (1987) found evidence of significant differentiation among aggregations in Kansas. The source of this differentiation, however, is not clear because the sample involved 4 nest aggregations along a 10-km section of the Kansas River and a fifth site 20-km distant on a separate tributary. It could be due to differentiation of the single most distant nest aggregation. Furthermore, the sample consisted of more than one female per nest making it difficult to determine realistic sample sizes for statistical analyses. In New York, I found no significant differentiation among nest aggregations along 5 of 6 streams (Kukuk *et al.* 1987b).

I used Crow and Aoki's (1984) method to estimate the product of group size (N) and the migration rate (m) among nest aggregations from G_{ST} values. G_{ST} values (which can be viewed as estimates of the probability that 2 alleles drawn randomly from a single nest aggregation are identical by descent) were calculated separately for 4 to 7 nest aggregations along each of six creeks. The solid line in FIGURE 1 illustrates the predictions of Crow and

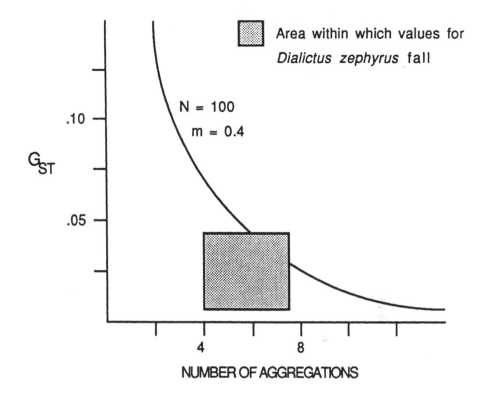

FIGURE 1. Adapted from FIGURE 2 of Crow and Aoki (1984), the line shows G_{ST} (a measure of population subdivision) for a given N (population size) and m (the proportion of genes exchanged per generation) in a habitat that is long and narrow. The gray area shows the region within which G_{ST} values lie for nest aggregations (4 to 7 each) along 6 creeks for *D. zephyrus* in central New York. Estimates of G_{ST} were calculated from the results of electrophoretic analyses of 8 enzyme coding loci of 40 individuals from each of 31 aggregations.

Aoki's model for finite populations arranged in a grid that is much longer than wide. Such an arrangement is similar to that found in nature for *D. zephyrus* , in which nest aggregations are located in a linear sequence along each creek. The locus by locus values of G_{ST} for 8 loci fall into a region on the graph corresponding to large Nm values. Populations along a single creek are neither small enough nor isolated enough to make a significant contribution to population structure (Kukuk *et al.* 1987b).

TABLE 1. Estimates of the coefficients of relationship (r) for aggregations of *D. zephyrus*. Wright's F_{ST} coefficient, and the frequency of the most common allele (f_{mca}) were used to estimate r using the formula of Aoki and Nozawa (1984):

$$r = 2F_{ST} / 1 + F_{ST}$$

LOCUS	F_{ST}	r	f_{mca}
Pgm	0.0169	0.0332	0.70*
Mdh-2	0.0083	0.0165	0.98
Pep-LA-2	0.0069	0.0137	0.88*
Me-2	0.0191	0.0375	0.96
Gda	0.0115	0.0227	0.98
Idh	0.0241	0.0471	0.98
Dia	0.1077	0.1940	0.98
Hbdh	0.0178	0.0350	0.66*
Adh	0.0210	0.0251	0.57*

Over all 9 loci, r = 0.046±0.018
Over 4 highly polymorphic loci(*), r = 0.027±0.005
Based on 7 aggregations, 280 individuals

Aoki and Nozowa (1984) use an analogous measure of population structure, Wright's F_{ST}, to calculate coefficients of relationship among groups. The results for nest aggregations along a single creek are shown in TABLE 1. The estimates using only highly polymorphic loci (those more likely to involve strictly neutral alleles) appear to be more conservative than for the less polymorphic loci. Even the higher values that are based on all loci in which the frequency of the most common allele is less that 0.99 indicate that there is insufficient genetic differentiation among nest aggregations for intra-aggregation relatedness to be important. The results for the 5 other creeks are similar (TABLE 2). Aoki and Nozawa (1984), reached a similar conclusion for human groups and suggested that an intergroup coefficient of relationship of from 0.1 to 0.15 was not sufficiently high for group selection to favor self-sacrificial altruism in primates.

TABLE 2. Coefficients of relationship, r (and standard deviations), within aggregations of *D. zephyrus* estimated using the method of Aoki and Nozawa (1984). Estimates are based on samples of approximately 40 females (one from each nest) from each of 31 aggregations along each of 6 creeks in central New York and northern Pennsylvania. Values of r_{all} were calculated using all allozyme loci for which the most common allele occurred at a frequency of less than 0.99 along a given creek. Values of r_{hp} were calculated with loci for which the most common allele occurred at a frequency less than 0.90.

CREEK	r_{all}	N_{all}	r_{hp}	N_{hp}
Six mile	0.047(0.056)	9	0.030(0.011)	4
Fall	0.027(0.013)	10	0.025(0.008)	4
Salmon	0.032(0.022)	10	0.030(0.010)	4
Pipe	0.027(0.024)	9	0.026(0.021)	4
Cayuta	0.034(0.031)	10	0.039(0.027)	4
Wappas.	0.031(0.016)	10	0.031(0.018)	4

The values for *D. zephyrus* aggregations are much less than those reported for primates.

TABLE 3 gives the results of a hierarchical analysis of electrophoretic data obtained from 37 nest aggregations located along 9 creeks in 3 drainage areas (see Kukuk *et al.* 1987b for a description of the sample) using the method of Weir and Cockerham (1984). The coancestries (analogous to G_{ST}) for drainages, rivers and sites are small, and suggest little differentiation among nest aggregations along creeks.

In combination with evidence for intra-aggregation population structure (reviewed below) these results suggest a dispersal pattern in which most individuals mate and "settle" near their natal nests while some individuals migrate. Migrants may

TABLE 3. Macrogeographic subdivision of *D. zephyrus* populations determined by hierarchial analyses using the method of Wier and Cockerham (1984). Samples used for analyses were from 37 aggregation sites, along 9 rivers, in 3 drainage areas in northern Pennsylvania and Central New York. Data from eight polymorphic loci were used in the analysis. The frequency of the most common allele for each locus ranged between 0.28 and 0.97. F is an estimate of the average inbreeding coefficient for these populations and is significantly greater than 0 ($p < 0.05$).

	Drainages	Rivers	Sites	F
Coancestries				
Estimates	0.0364	0.0517	0.0561	0.1040
Standard Devs.	0.0245	0.0364	0.0359	0.0329
Correlations				
Estimates	0.0156	0.0068	0.0498	
Standard Devs.	0.0131	0.0017	0.0126	

be as likely to move long distances as they are to move short distances (Crozier *et al.* 1987). Further investigation of dispersal in *D. zephyrus* is needed to confirm this conclusion.

GENETIC VARIATION AMONG NEIGHBORHOODS

Circumstantial evidence suggests the formation of related neighborhoods within nest aggregations, particularly in Kansas. Crozier *et al.* (1987) demonstrated microgeographic structure within two nest aggregations. They noted that both of these aggregations had suffered extinction and recolonization as recently as 4 years prior to collection of the samples. This genetic evidence is consistent with the behavioral findings of Kukuk and Decelles (1986) that suggested an intra-aggregation subdivision and demonstrated a clumped distribution of related colonies within a single aggregation in Kansas. Kukuk and Decelles (1986) hypothesized that this was due to female philopatry (females founding nests near their natal nest).

The pattern of annual colony initiation in a relatively large aggregation near Ithaca, New York, provides indirect support for this hypothesis that populations are subdivided into related neighborhoods. Early season colony initiations occur in clusters and not at random within aggregations (Kukuk unpublished data), suggesting that each cluster of foundresses consisted of the females from a single natal colony. A hierarchical analysis (Weir and Cockerham 1984) of large scale population structure of *D. zephyrus* in New York also suggests overall inbreeding (TABLE 3). Inbreeding could be due to subdivisions within nest aggregations which result in neighborhoods of related individuals in which mating preferentially occurs, although other interpretations are also possible. Consistent with this scenario is the discovery of diploid males in 2 of 3 nest aggregations in which males were examined electrophoretically (TABLE 4). This finding could also be due to founder effects causing the loss of sex determining alleles or small population sizes, however the later seems unlikely given the lack of genetic differentiation among nest aggregations (Kukuk and May submitted).

Finally, observations of male dispersal suggests that a substantial proportion of males may mate near their natal nests. Mating in *D. zephyrus* and many other halictine bees takes place within the nest aggregation; males patrol in search of receptive females (Batra 1966, Barrows 1976, Michener 1974, Smith and Ayasse 1987, Wcislo 1987). Neither territorial behavior nor male-male agonistic interactions have been observed. Males do not patrol the entire aggregation but remain in a limited area (cf. Smith and Ayasse 1987). This area appears to include a male's natal nest in a small *D. zephyrus* nest aggregation in New York (Kukuk in press).

This interesting, but preliminary, evidence suggests that *D. zephyrus* aggregations may be structured into neighborhoods of related colonies that may have some impact on the evolution of eusociality. This structure could be viewed as enhancing group selection, or as strengthening kin selection by increasing "population viscosity", which increases relatedness among interacting individuals. Additional studies of dispersal for both males and females are needed to confirm this hypothesis.

TABLE 4. Population structure and the occurrence of diploid males in three aggregations of *D. zephyrus*.

AGGREGATION	Robinson	Monkey Run	Salmon Creek-A
NUMBER OF MALES SAMPLED	60	125	24
NUMBER OF NESTS	200	1,000+	1,000+
AGGREGATIONS WITHIN 1 KM	none found	2 or more	2 or more
PROPORTION OF LOCI POLYMORPHIC	0.58*	0.42*	1.00*
ESTIMATED HETEROZYGOSITY	0.17 *	0.14 *	0.17*
OBSERVED HETEROZYGOSITY	0.18*	0.13*	0.16*
ESTIMATED FREQUENCY OF DIPLOID MALES	0.14 ($\pm$ 0.04)**	0.02 ($\pm$0.01)**	0.0

* Estimated by electrophoretic examination of the protein products of only 12 polymorphic loci (as described in Kukuk et al 1987) and should be considered for comparison among sites only (Nei 1978).
** Expected std. dev. in frequency for a binomial distribution, std. dev. = K_{pq}/k.

GENETIC VARIATION WITHIN COLONIES

Colony Structure

If the evolution of eusociality involves kin selection, or group selection in kin structured populations the genetic composition of colonies (social units) should entail high relatedness of beneficiaries to altruists. Colonies of social insects function as reproductive units in which the fitness of all members may be enhanced due to their cooperative interactions. In halictine bees the interaction structure (*sensu* Wade 1980) results from high levels of interactions within colonies versus low levels of interaction among individuals from different colonies. Assessement of the relatedness of cooperating versus competing individuals allows evaluation of the possible role of kin selection. Although the competition structure of *D. zephyrus* is not well known, it is probably reasonable to view all females within an aggregation as competitors. With this in mind, intracolony relatedness can be estimated using a sample of colonies collected from a single nest aggregation.

By definition, primitively eusocial colonies are composed of one singly mated female (the queen) and her daughters, who function as sterile workers that rear only the queen's offspring. In primitively eusocial halictine bees, such "ideal colonies" are not always found in nature (e.g., see Michener 1974, Breed 1975 1976, Eickwort 1985 1986, Packer and Knerer 1986, Sakagami 1980; but see also Plateaux-Quénu 1962). In general, deviations are due to multiple mating, as observed in *Halictus ligatus* (Packer, 1986b) or to multiple queens.

One form of multiple queen colonies is "serial polygyny," queen replacement by one of her daughters following the death of the foundress queen. Nest takeover after abandonment by the original foundress or nest usurpation are also possible. While queen replacement is well documented in *D. zephyrus* , other forms of serial polygyny are not well documented in halictine bees (but see Eickwort 1985) and deserve additional study. Serial polygyny may be most important for annual species in regions where the active season is short so that a complete turnover of workers prior to rearing reproductive brood is unlikely.

"Simultaneous polygyny" could occur via laying workers; in most species of primitively eusocial halictines some workers have developed ovaries, and some are inseminated (e.g., Breed 1976, Michener 1974, Packer and Knerer 1985). In addition, intraspecific cleptoparasitism could account for polygyny, as suggested for a subtropical population of *Halictus ligatus* (Packer 1986a).

Electrophoretic studies of colony composition have only been conducted for *D. zephyrus.* Colony composition can be inferred by inspection of allozyme phenohypes (Kukuk *et al.* 1987a) or through estimation of intracolony relatedness also based on electrophoretic data. Intracolony relatedness was calculated for adult females taken from mid-summer colonies in Kansas (Crozier *et al.* 1987) and for first brood, and second brood colonies in New York (Kukuk *et al.* unpublished data).

Genetic Variation in First Brood Workers

In this section I present results concerning genetic variation the first group of workers produced by queens in the spring in New York. This variation could be due to multiple mating of the queen or to serial polygyny. Electrophoretic phenotypes of each foundress and her female offspring in 31 apparent single foundress colonies of *D. zephyrus* indicate that 36% of these colonies contained worker progeny not accounted for by one singly mated queen. In 13% phenotypes of the brood could only by explained by more than one laying female while the remaining 23% could have been produced by one doubly mated female (Kukuk *et al.* 1987a).

It is not possible to easily estimate the proportion of multiply mated queens from these data as only a small number of females are produced in the first brood (Mean = 3.2) and sperm clumping or sampling error might give the appearance of single mating. The allozyme method confounds multiple maternity and multiple paternity as one cannot discern which parent contributed a particular allele.

The presence of two laying females in at least 13% of apparently singly founded *D. zephyrus* colonies could be the result of the takeover of abandoned nests, nest usurpation, intraspecific cleptoparasitism, or cryptic multiple foundresses. Many single foundresses die before the maturation of the first brood (Kukuk unpublished); it is possible that colonies could then be taken over by a second foundress. Successful nest usurpation has been observed in *D. zephyrus* on 2 occasions out of 7 attempts (Kukuk *et al.* 1987a). That it was seen at all suggests that nest usurpation is not a rare occurrence in *D. zephyrus.*

During the active foraging period a considerable number of adult females in a nest aggregation fly from nest entrance to nest entrance as if "lost". This behavior is cited often in the halictine literature and was interpreted by Packer (1986a) as attempted intraspecific cleptoparasitism. However, definitive proof that these females deposit eggs in nests is not available. Studies in which this behavior has been observed usually have been in aggregations disturbed by the investigator; these "lost" females may indeed be disoriented.

Average intracolony relatedness among the first brood (potential workers) was 0.57 (95% confidence limits of 0.35 and 0.78, estimated using the method of Queller and Goodnight (in press; TABLE 5). The first female brood are therefore probably not all super sisters.

Genetic Variation in Later Brood

Individuals reared later in the season may be the offspring of the queen or of workers. Intracolony relatednesses of *D. zephyrus* at the time of reproductive brood rearing in New York are also presented in TABLE 5. As with the first brood colonies, these are not strictly eusocial colonies. The average relatedness of workers (0.43 $\pm$0.10) is similar to the estimates for first brood. This suggests that the females making up the first generation of brood remain and become workers, i.e., individuals that are less related to other colony members remain. Females from the second brood (=gynes) appear to be more closely related, even though there is a great deal of overlap in the standard errors of the relatedness estimates. The relatedness of gynes from these colonies is close to that expected for super sisters (.75).

High costs of independent colony initiation can compensate for moderate levels of intracolony relatedness such as those in New York and altruism may still have a selective advantage. If an individual foundress can rear very few brood on her own or if she encounters high risks of both mortality and brood parasitism then the cost side of Hamilton's equation is affected (see Strassmann and Queller Chapter 5 this volume). Intracolony relatedness estimates for workers of *D. zephyrus* in Kansas are higher than those from New York (Crozier *et al.* 1987). The longer active season in Kansas may have yielded a sample of workers that were all the daughters of a surviving foundress or a replacement queen. In New York the workers may be the result of "serial polygyny" that occurs during the solitary phase of the colony cycle.

Mean relatedness of female brood to workers for colonies from two nest aggregations in New York (n = 9 and n = 10 respectively) fall between 0.26 and 0.61 (Kukuk and Crozier unpublished data).

TABLE 5. Nest aggregation site (Site), type of sample (Description), frequency of the most common allele (FMCA), number of colonies examined (n), estimates of intercolony relatedness (r; using the method of Queller and Goodnight, in press), standard errors (STE; calculated using the jacknife method), and overall 95% confidince limits for the Salmon Creek (SC, 1983) and Robinson (RO, 1985) nest aggregations in Central New York for the first brood (FB), workers, the second brood (SB), and all females from late summer colonies (EC).

Site	Des-cription	Locus	FMCA	n	r	STE	95% Conf. Limits
SC	FB	Pep-GL-2	0.71	15	0.50	0.17	
		Hbdh	0.67	"	0.67	0.16	
		Combined			0.57	0.11	0.35-0.78
RO	Workers	Pep-GL-2	0.83	11	0.76	0.10	
		Hbdh	0.32	"	0.29	0.08	
		Combined			0.43	0.10	0.23-0.63
	SB	Pep-GL-2	0.74	9	0.66	0.18	
		Hbdh	0.32	"	0.74	0.11	
		Combined			0.72	0.09	0.54-0.90
	EC	Pep-GL-2	0.78	12	0.59	0.16	
		Hbdh	0.32	"	0.50	0.10	
		Combined			0.53	0.10	0.33-0.73

SUMMARY

Genetic studies of *D. zephyrus* resolve some questions and raise others. Macrogeographic population structure suggests that nest aggregations are not genetically isolated from one another and may be short lived. There is circumstantial evidence to suggest that populations are structured within aggregations into neighborhoods of related colonies. Additional information on dispersal is needed to clarify the microgeographic population structure. The first brood in colonies is not strictly composed of the progeny of one, singly mated female but the second female brood may normally be super sisters. Detailed field studies documenting the proposed types of "serial polygyny" are needed along with application of new techniques to characterize the genotypes of individuals within colonies. The levels of relatedness in *D. zephyrus* colonies are high enough for kin selection to be important in the maintenance of worker altrusm given environmentally caused limitations on the reproductive output of solitary females.

Acknowledgments

The assistance of several colleagues was of great importance in carrying out much of this research. In particular I would like to acknowledge B. P. May, G. C. Eickwort and R. H. Crozier along with P. C. Decelles, D. C. Queller, D. R. R. Smith, and B. S. Weir. This manuscript was improved due to careful reading by B. P. May and G. C. Eickwort. Partial support for this work was provided by NSF grant #BSR-8407592.

LITERATURE CITED

Aoki, K. and K. Nozawa. 1984. Average coefficient of relationship within troups of the Japanese monkey and other primate species with reference to the possibility of group selection. *Primates* 25:171-184.

Barrows, E. M. 1976. Mating behavior in halictine bees (Hymenoptera: Halictidae): I. Patrolling and age-specific behavior in males. *J. Kansas Ent. Soc.* 49:105-119.

Batra, S. W. T. 1964. Behavior of the social bee, *Lasioglossum zephyrum*, within the nest. *Insectes Sociaux* 11:159-186.

Batra, S. W. T. 1966. The life cycle and behavior of the primitively social bee, *Lasioglossum zephyrum* (Halictidae). *Univ. of Kansas Sci. Bull.* 47:359-423.

Berkelhammer, R. C. 1982. Population genetic structure of dolichoderine ants. In *The biology of social insects*, M. D. Breed, C. D. Michener and H. E. Evans, eds. Westview Press:Boulder, CO.

Breed, M. D. 1975 Life cycle and behavior of a primitively social bee, *Lasioglossum rohweri* (Hymenoptera: Halictidae). *J. Kansas Entomol. Soc.* 48:64-80.

Breed, M. D. 1976. The evolution of social behavior in primitively social bees: a multivariate analysis. *Evolution* 30:234-240.

197

Breed, M. D. 1977. Interactions among individuals and queen replacement in a eusocial halictine bee. In *Proc. 8th International Congress IUSSI. Wageningen, Netherlands.*

Breed, M. D. and G. J. Gamboa. 1977. Behavioral control of workers by queens in primitively eusocial bees. *Science* 195:694-696.

Brothers, D. J. and C. D. Michener. 1974. Interactions in colonies of primitively social bees. III Ethometry of division of labor in *Lasioglossum zephyrum.* (Hymenoptera: Halictidae). *J. Comp. Physiol.* 90:129-168.

Buckle, G. R. 1982a. Queen-worker behavior and nestmate interactions in young colonies of *Lasioglossum zephyrum. Insectes Sociaux* 29:125-137.

Buckle, G. R. 1982b. Differentiation of queens and nestmate interactions in newly established colonies of *Lasiogossum zephyrum . Sociobiology* 7:8-20.

Buckle, G. R. 1984. A second look at queen-forager interactions in the primitively eusocial halictid, *Lasioglossum zephyrum. J. Kansas Ent. Soc.* 57:1-6.

Buckle, G. R. 1985. Increased queen-like behaviour of workers in large colonies of the sweat bee *Lasioglossum zephyrum. Anim. Behav.* 33:1275-1280.

Craig, R. and R. H. Crozier. 1979. Relatedness in the polygynous ant *Myrmecia pilosula. Evolution* 33:335-341.

Crow, J. F. and K. Aoki. 1984. Group selection for a polygenic behavioral trait: Estimating the degree of population subdivision. *Proc. Nat. Acad. Sci. USA* 81:6073-6077.

Crozier, R. H. 1977. Evolutionary genetics of the Hymenoptera. *Ann. Rev. Entom.* 22:263-288.

Crozier, R. H. 1980. Genetical structure of social insect populations. In *Evolution of social behavior: hypotheses and empirical tests,* H. Markl, ed. Dahlem Konferenzen:Weinheim.

Crozier, R. H., P. Pamilo and Y.C. Crozier. 1984. Relatedness and microgeographic genetic variation in *Rhytidoponera mayri,* an Australian arid-zone ant. *Behav. Ecol. Sociobiol.* 15:143-150.

Crozier, R. H, B. H. Smith and Y. C. Crozier. 1987. Relatedness and population structure of the primitively eusocial bee *Lasioglossum zephyrum* (Hymenoptera: Halictidae) in Kansas. *Evolution* 41:902-910.

Eickwort, G. C. 1985. The nesting biology of the sweat bee *Halictus farinosus* in California with notes on *H. ligatus* (Hymenoptera: Halictidae). *Pan-Pac. Entomol.* 61:122-127.

Eickwort, G. C. 1986. First steps into eusociality: The sweat bee *Dialictus lineatulus. Fla. Entomol.* 69:742-754

Eickwort, G. C. and P. F. Kukuk. 1987. Reproductive castes in primitively eusocial halictid bees. In *Chemistry and Biology of Social Insects,* J. Eder and H. Rembold, eds. Verlag J. Peperny:Munich. pp. 261-262.

Fix, A. G. 1985. Evolution of altruism in kin-structured and random subdivided populations. *Evolution* 39:928-939.

Grafen, A. 1986. Split sex ratios and the evolutionary origins of eusociality. *J. theor. Biol.* 122:95-121.

Greenberg, L. and G. R. Buckle. 1981. Inhibiting of worker mating by queens in a sweat bee (*Lasioglossum zephyrum*). *Insectes Sociaux* 28:347-352.

Halliday, R. B. 1983. Social organization of meat ants *Iridomyrmex purpureus* analyzed by gel electrophoresis of enzymes. *Insectes Sociaux* 30:45-56.

Hamilton, W. D. 1963. The evolution of altruistic behavior. *Am. Nat.* 97:354-356.

Hamilton, W. D. 1964. The genetical evolution of social behavior, I and II. *J. theor. Biol.* 7:1-52.

Hamilton, W. D. 1972. Altruism and related phenomena, mainly in social insects. *Ann. Rev. Ecol. Syst.* 3:192-232.

Kukuk, P. F. Dispersal of males of the primitively eusocial sweat bee *Dialictus zephyrus* within a small nest aggregation. *Sociobiology* in press.

Kukuk, P. F. and P. C. Decelles. 1986. Behavioral evidence for population structure in *Lasioglossum (Dialictus) zephyrum* female dispersion patterns. *Behav. Ecol. Sociobiol.* 19:233-239

Kukuk, P. F. and B. P. May. Dominance hierarchy in the primitively eusocial bee *Lasioglossum (Dialictus) zephyrum*: Is genealogical relationship important? *Anim. Behav.* in press.

Kukuk, P. F. and B. P. May. The occurrence of diploid males in a primitively eusocial halictine bee *Lasioglossum (Dialictus) zephyrum.* Submitted.

Kukuk, P. F., G. C. Eickwort and B. P. May. 1987a. Multiple maternity and multiple paternity in first generation brood from single foundress colonies of the sweat bee *Dialictus zephyrus.* *Insectes Sociaux* 34:131-135.

Kukuk, P. F., R. H. Crozier, D. R. R. Smith, G. C. Eickwort and B. P. May. 1987b. The population structure of a primitively eusocial sweat bee, *Lasioglossum (Dialictus) zephyrum* . In *Chemistry and Biology of Social Insects,* J. Eder and H. Rembold, eds. Verlag J. Peperny:Munich p. 364.

Lester, L. J. and R. K Selander. 1981. Genetic relatedness and the social organization of *Polistes* colonies. *Am. Nat.* 117:147-166.

Metcalf, R. A. 1980. Sex ratios, parent-offspring conflict, and local competition for mates in the social wasps *Polistes metricus* and *Polistes variatus*. *Am. Nat.* 166:642-654.

Metcalf, R. A. and G. S. Whitt. 1977. Intra-nest relatedness in the social wasp *Polistes metricus*. *Behav. Ecol. Sociobiol.* 2:339-351.

Michener, C. D. 1974. *The social behavior of the bees. A comparative study.* Belknap Press of Harvard Univ. Press:Cambridge, Mass.

Michener, C. D. and D. J. Brothers. 1974. Were workers of eusocial Hymenoptera initially altruistic or oppressed? *Proc. Nat. Acad. Sci. USA* 71:671-674.

Michener, C. D., D. J. Brothers and D. R. Kamm. 1971a. Interactions in colonies of primitively social bees. II: Artificial colonies of *Lasioglossum zephyrum*. *Proc. Nat. Acad. Sci. USA* 68:1241-1245.

Michener, C. D., D. J. Brothers and D. R. Kamm. 1971b. Interactions in colonies of primitively social bees. II. Queen-worker relations in *Lasioglossum zephyrum*. *J. Kansas Ent. Soc.* 44:276-279.

Michod, R. E. 1982. The theory of kin selection. *Ann. Rev. Ecol. Syst.* 13:23-55.

Michod, R. E. and R. Abugov. 1980. Adaptive topography in family-structured models of kin selection. *Science* 210:667-669.

Nei, M. 1978. Estimation of average heterozygosity and genetic distance from a small number of individuals. *Genetics* 89:583-590.

Nunney, L. 1985. Group selection, altruism and structured-deme models. *Am. Nat.* 126 212-230.

Packer, L. 1986a. The biology of a subtropical population of *Halictus ligatus* IV: A cuckoo-like caste. *J. New York Entomol. Soc.* 94:458-466.

Packer, L. 1986b. The social organization of *Halictus ligatus* (Hymenoptera: Halictidae) in southern Ontario. *Can. J. Zool.* 64:2317-2324.

Packer, L. 1987a. Competition over oviposition in two populations of *Halictus ligatus*. In *Chemistry and Biology of Social Insects*, J. Eder and H. Rembold, eds. Verlag J. Peperny:Munich. pp. 553-554.

Packer, L. and Knerer, G. 1985. Social evolution and its correlates in bees of the subgenus *Evylaeus* (Hymenoptera: Halictidae). *Behav. Ecol. Sociobiol.* 17:143-149.

Packer, L. and Knerer, G. 1986. The biology of a subtropical population of *H. ligatus* Say (Hymenoptera: Halictidae) I. Phenology and social organization. *Behav. Ecol. Sociobiol.* 18:636-375.

Pamilo, P. 1981. Genetic organization of *Formica sanguinea* populations. *Behav. Ecol. Sociobiol.* 9:45-50.

Pamilo, P. 1982. Genetic population structure in polygynous *Formica* ants. *Heredity* 48:95-50.

Pamilo, P. 1983. Genetic differentiation within subdivided populations of *Formica* ants. *Evolution* 37:1010-1022.

Pamilo, P. 1984. Genotype correlation and regression in social groups: multiple alleles, multiple loci and subdivided populations. *Genetics* 107:307-320.

Pamilo, P., and R. Rosengren. 1984. Evolution of nesting strategies of ants: Genetic evidence from different population types of *Formica* ants. *Biol. J. Linn. Soc.* 21:311-348.

Pamilo, P., and S. Varvio-Aho. 1979. Genetic structure of nests in the ant *Formica sanguinea. Behav. Ecol. Sociobiol.* 6:91-98.

Pearson, B. 1982. Relatedness of normal queens (macrogynes) in nests of the polygynous ant *Mrymecia rubra* Latreille. *Evolution* 36:107-102.

Pearson, B. 1983. Intra-colonial relatedness amongst workers in a population of nests of the polygynous ant, *Myrmica rubra* Latreille. *Behav. Ecol. Sociobiol.* 12:91-98.

Plateaux-Quénu, C. 1962. Biology of *Halictus marginatus* Brullé. *J. Apic. Res.* 1:41-51.

Queller, D. C. and K. F. Goodnight. Estimation of genetic relatedness using allozyme data. *Evolution* in press.

Sakagami, S. F. 1980. Bionomics of the halictine bees in northern Japan I.*Halictus (Halictus) tsingtouensis* (Hymenoptera: Halictidae), with notes on the number of origins of eusociality. *Kontyû* 48:526-536.

Smith, B. H. 1987. Effects of genealogical relationship and colony age on the dominance hierarchy in the primitively eusocial bee *Lasioglossum zephyrum. Anim. Behav.* 35:211-217.

Smith, B. H., and M. Ayasse. 1987. The role of kin recognition in the mating preferences of male halictine bees (Hymenoptera: Halictidae). In *Chemistry and Biology of Social Insects,* J. Eder and H. Rembold, eds. Verlag J. Peperny:Munich p. 470.

Taylor, P. D. and D. S. Wilson. 1988. A mathematical model for altruism in haystacks. *Evolution* 42:193-196.

Unoyama, M. K., and M. W. Feldman. 1980. On relatedness and adaptive topography in kin selection. *Theor. Pop. Biol.* 19:87-123.

Wade, M. J. 1980. Kin selection: its components. *Science* 210:665-667.

Ward, P. S. and R. W. Taylor. 1981. Allozyme variation, colony structure and genetic relatedness in the primitive ant *Nothomyrmecia macrops* Clark (Hymenoptera: Formicidae). *J. Aust. Entomol. Soc.* 20:117-183.

Wcislo. W. 1987. The role of learning in the mating biology of a sweat bee *Lasioglossum zephyrum* (Hymenoptera: Halictidae). *Behav. Ecol. Sociobiol.* 20:179-185.

Weir, B. S. and C. C. Cockerham. 1984. Estimating F statistics for the analysis of population structure. *Evolution* 38:1358-1370.

Wilson, D. S. 1977. Structured demes and the evolution of group-advantageous traits. Am. Nat. 111:157-185.

Wilson, D. S. 1980. *The natural selection of populations and communities.* Benjamin-Cummings:Menlo Park, CA.

Wilson, D. S. 1983. The group selection controversy: History and current status. *Ann. Rev. Ecol. Syst.* 14:159-187.

Wilson, D. S. 1987. Altruism in Mendelian populations derived from sibling groups: The haystack model revisited. *Evolution* 41:1059-1071.

Wilson, E. O. 1971. *The insect societies.* Belknap Press of Harvard University Press:Cambridge, Mass.

INDEX

foraging 9, 17, 18, 19, 24, 33, 37, 46, 61, 62, 63, 64, 65, 66, 67, 68, 71, 72, 89, 115, 164, 166, 194
    nectar forager 24, 63, 64, 65, 66, 67, 68, 71, 72, 164
    pollen forager 17, 18, 19, 24, 64, 65, 66, 67, 68, 71, 72
foreign-label-rejection model *see* kin recognition
Formicidae 123, 126, 131
Formica 125, 126
    rufa 126
foundress, founder 2, 8, 31, 32, 46, 52, 53, 54, 55, 88, 92, 103, 104, 106, 109, 111-116, 123, 133, 136, 164, 169, 173, 174, 180, 184, 190, 191, 193, 194, 195
full sister 3
    see also genetic relatedness
gamergate 135
genetic
    additive variance 20, 23, 37, 38, 40, 41, 42, 43, 45, 49, 50, 71
    allelic models 9, 25, 33, 34, 36, 42, 45, 49, 50, 108, 124, 153, 154, 168, 170, 174-179, 180, 186, 188
    allelomorph 63, 108
    allozyme 33, 63, 64, 70, 105-106, 108, 125-126, 139, 189, 193, 194
    diplodiploid 2-3
    diploid 2, 3, 5, 83, 84, 124, 125, 170, 191, 192
    dominance variance 50-51
    drift 22
    haplodiploid 2, 3, 5, 36, 38, 39, 42, 82-85, 104, 105, 124
    haploid 2, 3, 17, 131, 152, 168, 170
    Hardy-Weinberg equilibrium 25, 35, 39, 40, 42, 46
    heterogeneity 16, 22
    inbreeding 3, 5, 126, 132, 167, 180, 190, 191
    intracolonial variation 70, 73
    markers 64, 70, 150
    non-additive variance 20, 23, 37, 40, 41, 42, 50, 71
    pairing terminology 3
    pleiotropic 9
    polymorphism 23, 25
    quantitative 32, 48, 51, 180
    regulation 62
    relatedness 2, 3, 4, 5, 6, 15, 50, 62, 82-86, 93-94, 103, 104, 105-110, 111, 116, 117, 150, 154, 159, 167, 168, 169, 185-197
    sex linked 39

specialists 16, 24-25, Chapter 4
structure 2, 4, 15, 27, 61- 63, 70
subfamily 4, 17, 20-21, Chapter 4
variation 3, 8, 16-23, 27, 33, 48, 50, 55, 61-62, 70, 73,
75, 130, 132, 135, 138, 139, 153, 169, 186, 188, 190-195
see also chromosome, fitness, genome, phenotype
genitalia 131, 132, 139
genome 2, 3, 71
see also genetic
geographic distributions *see* allopatric, peripatric, sympatric
geographical variation 124, 125-130, 132-133, 138, 139, 190, 196
gestalt model *see* kin recognition
group selection *see* selection
group size 86, 87, 88, 91, 93, 111
guard, guarding *see* defense
gyne 81, 126, 134, 135, 138, 140, 184, 195
macrogyne 126, 134, 138, 140
microgyne 126, 134, 138, 140
see also female, monogynous, polygynous, uninseminated
queens, queen
Halictus ligatus 193
Halictinae 183, 185, 191, 193, 194
Hamilton (William D.) 5, 7, 9, 82-85, 103-104, 106-107, 111, 115,
185, 195
haplodiploid *see* genetic
haploid *see* genetic
Hardy-Weinberg equilibrium *see* genetic
helping behavior 7, 8, 81-88, 104, 117
see also altruism
heritability 48, 50-51
heterochromatin *see* chromosome
heterosis 41
honey bee 2, 3, 4, 6, 9, Chapter 2, 33, 34, 47, 48, 51, Chapter 4
see also Apis mellifera
hormone 70,112
juvenile 73
hybrid zone 126
hybridization 126
hygienic behavior 18, 64, 70
inbreeding *see* genetic
inclusive fitness *see* fitness
infertility 10
inquiline *see* parasite

Printed and bound by CPI Group (UK) Ltd, Croydon, CR0 4YY

23/10/2024

01778244-0008